science
and
technology

AN INTRODUCTION TO THE LITERATURE

DENIS GROGAN

BA FLA

HEAD OF THE DEPARTMENT OF BIBLIOGRAPHICAL STUDIES

COLLEGE OF LIBRARIANSHIP WALES

CLIVE BINGLEY  LONDON

FIRST PUBLISHED 1970 BY CLIVE BINGLEY LTD
16 PEMBRIDGE ROAD LONDON WII
SET IN 10 ON 12 POINT LINOTYPE PLANTIN
AND PRINTED IN GREAT BRITAIN BY THE CENTRAL PRESS (ABERDEEN) LTD
COPYRIGHT © DENIS JOSEPH GROGAN 1970
85157 095 X

CONTENTS

INTRODUCTION

This is a book written primarily for students, not practitioners. For the toiler actually in the field of scientific and technical information there are a host of excellent guides, as chapter 2 will indicate: this work is for the *would-be* practitioner, who has reached an understanding of the the general sources of information such as encyclopedias and yearbooks and bibliographies, but now wishes to move on to the sources of scientific and technical information. Ultimately, in the field (or even in the classroom or seminar), he may well be required to concentrate his attention on a specific subject area within science or technology. It is hoped that this text will not only initiate him into the structure of the literature of science and technology, but will also prepare the way for just that kind of detailed study of constituent subject parts.

Of course, many experienced scientific and technical information workers were obliged to commence their study of the literature of their subject at the deep end, so to speak, inasmuch as they had to learn on the job, working with the literature and its users in an appropriate specialised library. And for a librarian to master the literature of a particular subject, such daily use is still the best and probably the only satisfactory method. But this haphazard approach is not only unscientific, it is inefficient: it is true that the requisite familiarity with the literature can be attained this way, but it takes far longer than it needs.

The information workers of tomorrow, on the other hand, are given the opportunity to embark on the study of the literature in a more systematic fashion. As the students of today, not only do they commence with a basic course in general reference sources before progressing to scientific and technical literature, but they are encouraged to investigate the general structure of that literature before turning to more specific areas within the field. As novice practitioners in the library such students will of course still be faced with the need to come to grips with the detailed literature of their subject, but the hope is that they will be able to accomplish this more methodically and rapidly by virtue of their theoretical grasp of the overall pattern of the literature of science and technology. Perhaps even more importantly, such students should be particularly well placed to meet the challenge that will increasingly confront the future practitioner in all fields: rapid obsolescence of current knowledge as a result of accelerating technical advance.

It is hoped that there will be at least some practising librarians or information workers for whom this book will have an interest: those for example thoroughly familiar with their own special area, but who might appreciate the opportunity to take a more general view, or those contemplating or already embarked upon a change of subject field, particularly from the humanities to the sciences. Here I have been fortunate enough to be able to draw on my own experience over several years in supervising the day-to-day work of public service staff in a very large scientific and technical library.

The needs of the reader without formal scientific training have been kept especially in mind. A survey[1] carried out in 1965 indicated that over a third of the staff employed in scientific and technical information work in the United Kingdom fell into this category, and the ' swing from science ' in the schools and universities suggests that this proportion will increase.[2] For such students (and indeed for such practitioners) it is of particular importance that they have the opportunity of examining the general structure of the literature of science and technology before exploring an area in depth.

It is popularly thought that the way a librarian studies the literature of a subject is to memorise as many authors and titles as possible. Ever since the library catalogue was invented, this has never been really necessary, but even in these days of vast and instantaneous computer memories, many (including myself) have been reluctant to shed their admiration for the traditional reference librarian's encyclopedic knowledge of his stock, and with Goldsmith's village rustics we still marvel ' that one small head could carry all he knew '.

[1] A P J Edwards 'A national survey of staff employed on scientific and technical information work ' *Journal of documentation* 22 1966 210-244.

[2] *Enquiry into the flow of candidates in science and technology into higher education* (HMSO 1968) (Cmnd 3541).

However, a guide to the literature for a student demands a radically different approach from a guide to the literature for the user in the library. The latter is searching for information, and wants his guide to be comprehensive; the former is striving to understand the underlying pattern of the literature and needs a guide that demonstrates this by pointing to particular selected examples worthy of closer study. The object of attention is the general rather than the particular, or more precisely, it is the particular not for its own sake but only insofar as it represents an instance of the general. The zoology student dissects the dogfish not because it is the most important or even the most common of the fishes but because it serves particularly well as a representative of one of the group's two main types.

The wide selection of worthy books enumerating and describing those scientific and technical reference books and bibliographies that librarians and others find useful makes it unnecessary here even to attempt to list basic titles. This is a teaching tool, and as the essence of teaching is selection, attention is concentrated on *types* of literature, and individual titles are normally listed merely as manifestations of those types. The student should of course endeavour to examine them closely as representatives of their class, but since they are merely examples chosen from many possible alternatives, he might gain an even more valuable insight into this particular aspect of scientific and technological communication if he identifies for himself in the collections to which he has access similar instances of each type, and devotes his attention to these. Where a guide such as this has to endeavour to be comprehensive is in demonstrating all the *types* of scientific and technical literature a student might meet. Of course, once in the field, his task will be to build on this largely theoretical framework by identifying within each type no longer simply representative titles, but *all* those works within his chosen area which might be of value to their potential users in his library.

With Walford[3] already in possession of the field, and, one hopes, by every student's elbow, and the latest Winchell[4], greatly strengthened in the pure and applied sciences chapters, it would be superfluous in this book to rehearse the features of many of the titles quoted as examples. However, both of these guides concentrate on reference

---

[3] A J Walford *Guide to reference material: volume I, Science and technology* (Library Association, second edition 1966).

[4] C M Winchell *Guide to reference books* (Chicago, American Library Association, eighth edition 1967) and *Supplement* (1968).

works, whereas much of the actual literature used by the scientist and technologist is non-reference material such as monographs and textbooks, as well as a whole range of non-book materials such as periodicals, patents, research reports, etc. Obviously comment in the text on examples quoted from these categories is more appropriate, as it is with certain other groups of reference material such as British directories and bibliographies of scientists (excluded by Walford) and reviews of progress (excluded by Winchell).

It should be emphasised that this is a guide to the *literature,* and it is a mistake to assume that this is synonymous with information. It is an even graver error to underestimate the importance of non-documentary information in science and technology: a number of surveys have shown that 'live' sources (eg, consultation with colleagues, attendance at professional meetings, etc) play a large part in communication. And quite apart from any other reason, the pace of development is such that in many fields any information that has got into print has almost certainly been overtaken by events. It has been rightly said that the main disadvantage of the literature is that it is history and not news. Nevertheless, it is equally foolish to sell short the printed word: this is to throw away cumulative human experience. In Carlyle's words: 'All that mankind has done, thought, gained or been—it is lying, as in magic preservation, in the pages of books'. Surveys have shown a surprising amount of duplicated research and general wasted effort due to disregard of the published literature. Its vital role has been highlighted in recent years by the mushrooming computer-based information systems. These, of course, are 'literature-based', in that input and output are both ultimately in the form of printed words.

The reader will not find here any advice on how to use libraries and library catalogues, or any account of classification schemes. No attempt will be made to instruct him in the answering of reference enquiries or in the compilation of bibliographies. It is true that such assistance is given in many guides to the literature, and they are useful and indeed indispensable accomplishments of the 'compleat' librarian. This book, however, is written in the belief that personal service to the users of our libraries can best be improved by a more sophisticated approach to the literature by the librarian, allied to a more refined awareness of the user's needs. A much deeper understanding is required of the types of literature and of the special role of each type in the network of scientific communication. Of course the literature is only part of the pattern, but it is that part peculiarly within the librarian's domain and he should understand it fully. Merely to match subjects is to operate far below the optimum level of service. To provide 'something on' the topic the user is interested in is not enough. Librarians have much to learn here from the scientific and technological publishers, commercial and otherwise: they do not just produce 'books' and 'periodicals' on a subject. They (and their authors) direct their productions at particular groups of consumers, personal and institutional: although somewhat similar to the casual glance, textbooks are really quite different from monographs, research journals quite different from technical journals.

The student will find that practically all of the examples chosen are in the English language: he will appreciate that this is a mere matter of convenience, and that in many subjects there are vital works available only in other tongues. Again, for reasons of convenience, if no place of publication is cited with the publisher's name, it means that London appears in the imprint. Medicine has not been rigorously excluded: where a point can best be illustrated by a medical example it has been used. The lists of further reading appended to the chapters are deliberately highly selective, being confined to items thought to be of real value to the student, not too inaccessible, and capable of being read during a course of study. No attempt has been made to document every reference and quotation in the text: such excess of bibliographical scruple is out of place in a textbook for students.

It has been assumed throughout that the user of this book is familiar with *general* reference and bibliographical sources and the relevant terminology. He will not find, therefore, his attention drawn in the chapter on biographical sources to *Who's who* or the *Dictionary of national biography,* even though both works contain their share of scientists and technologists. *The British union-catalogue of periodicals* is not mentioned among the lists of periodicals for the same reason, although it contains the locations of thousands of scientific journals. And so on.

I owe a great debt to the authors of those classic guides to the literature of the various sciences and technologies, some of which are noted in chapter 2. They have been constantly consulted over the years during my own exploration of the field. It should be added, however, that practically all the works mentioned in the text I have personally examined, and in most cases made use of.

Aberystwyth   D J GROGAN
February 1970

I

THE LITERATURE

The greatest contribution that science has made to human progress has been the discovery and perfecting of the experimental method. Indeed, several writers have claimed that science is no more than this method —the scientific method as it is sometimes called. This method holds true for all the sciences, and the technologies also, and is of course widely applied in other disciplines.

The implications for the literature of science and technology are so far-reaching that it is essential for the student librarian to grasp the elements of the method. The first step a scientist (or technologist) takes towards solving a problem is to collect all the information that may have a bearing on the question: this is the *observation* stage. He then formulates a tentative theory as to how such facts are to be interpreted: this is the *hypothesis* stage. He then designs and carries out a series of controlled tests to try to confirm his working hypothesis: this is the *experimental* stage. If findings of the experiments prove his theory correct he formulates his answer to the problem: this is the *conclusion* stage. Of course, it frequently happens that the working hypothesis does not stand up under experiment, and the scientist must go back as often as necessary until he achieves a hypothesis that not only accounts for all the observed facts but can be confirmed by controlled experiment.

It is easy to see why science has been defined as ' tested knowledge '. If science progresses by the accumulation of facts derived from observation and experiment, then it is clearly vital for the scientist wishing to advance knowledge in his field to know what has already been achieved. He turns therefore to those records of observations and experiments left by his predecessors, to the literature, in fact. As even Sir Isaac Newton acknowledged: ' If I have seen further than most men it is by standing on the shoulders of giants '.

THE PRIMARY SOURCES

The original reports of scientific and technological investigations make up the bulk of what is known as the *primary* literature. Some of these records may be largely observational (*eg*, reports of scientific expeditions), or descriptive (*eg*, some trade literature), but most of them are accounts of experiments with findings and conclusions. It is a basic principle of scientific investigation that sufficient detail should be given to enable the work described to be repeated (and therefore double-checked) by any competent investigator.

These contributions then represent new knowledge (or at least new interpretations of old knowledge) and constitute the latest available information. They are published in a variety of forms:

1 Periodicals (many of these are solely devoted to reporting original work)
2 Research reports
3 Conference proceedings
4 Reports of scientific expeditions
5 Official publications
6 Patents
7 Standards
8 Trade literature
9 Theses and dissertations.

Many of course remain unpublished, and outside the mainstream of scientific progress, but do occasionally become accessible later in their original form, and are often consulted for their historical interest, *eg*,

1 Laboratory notebooks, diaries, memoranda, etc
2 Internal research reports, company files, etc
3 Correspondence, personal files, etc.

By its very nature the primary literature is widely scattered, disconnected, and unorganised. It records information as yet unassimilated to the body of scientific and technological knowledge. Although of vital importance, it is difficult to locate and to apply, and over a period there has therefore grown up a second tier of more accessible information sources.

SECONDARY SOURCES

These are compiled from the primary sources and are arranged according to some definite plan. They represent ' worked-over ' knowledge rather than new knowledge, and they organise the primary literature in more convenient form. By their nature they are often more widely available than the primary sources, and in many cases more self-sufficient:

1 Periodicals (a number of these specialise in interpreting and commenting on developments reported in the primary literature)
2 Indexing and abstracting services
3 Reviews of progress
4 Reference books, *eg,*
    a) encyclopedias
    b) dictionaries
    c) handbooks
    d) tables
    e) formularies
5 Treatises
6 Monographs
7 Textbooks.

In addition to repackaging the information from the primary literature many of these have the further useful function of guiding the worker to the original documents. In other words, they serve not only as repositories of digested data, but as bibliographical keys to the primary sources.

TERTIARY SOURCES
It is possible to distinguish a less well-defined group of sources the main function of which is to aid the searcher in using the primary and secondary sources. They are unusual in that most of them do not carry ' subject ' knowledge at all:

1 Yearbooks and directories
2 Bibliographies, *eg,*
    a) lists of books
    b) location lists of periodicals
    c) lists of indexing and abstracting services
3 Guides to ' the literature '
4 Lists of research in progress
5 Guides to libraries and sources of information
6 Guides to organisations.

NON-DOCUMENTARY SOURCES
These form a substantial part of the communication system in some disciplines within science and technology, as user surveys have shown.

It is clear that they provide something that the other sources do not (and perhaps cannot):

1 Formal, *eg,*
    a) government departments, central and local
    b) research organisations
    c) learned and professional societies
    d) industry, private and public
    e) universities and colleges
    f) consultants

2 Informal, *eg,*
    a) conversations with colleagues, visitors, etc
    b) ' corridor meetings ' at conferences, etc.

In many cases of course these oral sources serve merely as pointers to the documentary (primary or secondary) sources, and the following up of a ' personal contact ' often leads ultimately to the printed page.

USE OF THE LITERATURE

The demands of the experimental method being what they are, it is obvious that scientists and technologists are in constant need of information. Now there are three ways to find out a fact that you need: you can work it out for yourself, you can ask someone who knows, or you can look it up. We tend to avoid the first where we can, even if it is quite feasible, such as measuring the boiling point of alcohol or the weight of a seagull. The second way, by asking someone, is often the easiest, and for certain kinds of information an important channel of scientific and technological communication. It is the third way, however, that concerns us here, ' looking it up ' by consulting the literature.

A worker may consult all these types of sources for a host of reasons, but in general he uses literature only when he finds it essential for some particular facet of his task. It is easier to understand the pattern of use (and indeed the pattern of the information sources) if one groups various kinds of consultation into a series of ' approaches ', as has been demonstrated by Melvin J Voigt.[1]

a) *The current approach* arises from the need to keep up, to know what other workers in the field are doing (or about to do). Surveys of both scientists and technologists have shown that the most important source of such information is a fairly small core of primary

[1] M J Voigt *Scientists' approaches to information* ([Chicago] American Library Association, 1961).

periodicals, systematically scanned. Of great importance too are personal contacts with colleagues, with visitors, and at conferences. Secondary sources such as abstracts and indexes, if used at all, are regarded as guides to the primary sources.

Outside a worker's specific field of interest, he often tries to keep up to date in a broader area, knowledge of which is necessary to give depth and meaning to his special subject. Here it is not necessary for him to know his material so intimately or thoroughly, nor so vital to be absolutely up to the minute, and he can place more reliance on secondary sources. Abstracts are systematically perused as substitutes for reading the original papers. Reviews of progress are of particular value, and monographs (often located *via* book reviews in periodicals) are found useful for updating, despite their lateness in appearing.

b) *The everyday approach* arises in the course of daily work, regularly and frequently, usually in the form of a need for some specific piece of information vital for further progress. Unlike information needed in the current approach, it can usually be found in a number of places, and to the user the source is of little concern provided it is reliable. Almost invariably the one nearest to hand is chosen, and the emphasis is on finding what is wanted as soon as possible. This approach is the one most frequently used by scientists and technologists.

Probably because it is easiest, consulting other people is the most common method: it is probably more important here than all the printed sources put together, though it should be pointed out again that such consultation often leads to the literature. It is of course the secondary sources that are most commonly used. Most of them have been specifically compiled and arranged to furnish just such everyday information, *eg*, encyclopedias, dictionaries, handbooks, tables, etc, and they usually provide what the searcher wants without making him look beyond the book in hand. Bibliographical sources are less important for this everyday approach: not only do they take longer to peruse but they usually provide more on a topic than is wanted. And in many cases they do not provide actual subject data, merely indicating references. Because of their unorganised state, primary sources are used as the last resort, unless the searcher knows exactly where to look.

A distinct variation of the everyday approach might be called *The background approach*. This too arises in the course of daily work, but in the form of a demand not for a specific piece of information, nor even for ' something on ' a topic, but for an outline account sufficient to enable the worker to understand a subject new or unfamiliar to him. This has been called the ' orientation ' or ' brush up on a field ' approach, but J D Bernal's ' pilot information ' is as expressive a description as any. According to Bernal this need is ' fairly infrequent in fundamental research but extremely common in applied research where projects are often embarked on in fields quite unfamiliar to the research teams '. This approach is common too with teachers (and students) who often do not have the time to digest original material. The secondary sources most sought after for this purpose are reviews of progress: a recent ' state-of-the-art ' article on the required topic is usually the ideal answer.

c) *The exhaustive approach* usually arises when work begins on a new investigation, and involves a check through all the relevant information on a given subject. It is called for less frequently than the current or everyday approaches, but is vitally important, and often urgent. ' Exhaustive ' is almost always a relative term: E J Crane, for many years Director and Editor of *Chemical abstracts*, has written ' it is necessary to learn to recognize the point beyond which it is not reasonable to go because of the improbability of finding something of sufficient value to compensate for the time, effort, and expense of proceeding further '.

This approach is dependent very largely on printed sources, and bibliographical tools such as indexes and abstracts are all-important, but it must not be forgotten that literature references are included in many other secondary sources, *eg*, treatises, monographs, encyclopedias, etc. Identifying the primary sources is of course the object of the literature search, but once they have been located they can then themselves be used in the search by following the references invariably cited at the end of every original article. This, picturesquely called the ' snowball method ', is a very common way of proceeding.

d) *The browsing approach*, by definition unplanned, is nevertheless a fruitful path to information, definitely part of the scientific and technological communication system, and could well be added to Voigt's three approaches. Here, deliberately or otherwise, but without any specific topic in mind, workers seek information outside their predefined areas of interest, the aim (or result) being to extend the boundaries of that interest. Surveys have shown that printed sources are particularly productive in stimulating new ideas and interests.

Much of what we know about the use of the literature of science and technology is derived from ' user surveys ', of which we now have a very large number. It is not possible here even to summarise the findings,[2] and in any case it is unwise to draw conclusions as to the *value* of different types of literature solely on the basis of surveys of *use*, but it is clear that there is a surprising uniformity of literature practices within particular subject fields. Between disciplines, however, there are marked variations: pure scientists for example are far more dependent on the literature than industrial scientists and technologists. Even within a particular discipline it is often possible to chart variations according to type of activity: in engineering, predictably perhaps, 'pure' or 'academic' literature (primary journals, research reports, reviews of progress, etc) is used predominantly by those in research and teaching; 'applied' or 'industrial' literature (handbooks, standards, trade journals, etc) is used mainly by those whose work is concerned with design, maintenance, and testing.

Paradoxically, the most striking finding of these surveys is that scientists and technologists do not use the literature enough. In an ASLIB survey published in 1964, 647 research scientists reported 43 instances of finding literature, too late, which showed that their own research duplicated other work, and 245 instances of finding literature too late for the information to have full value. In a 1966 survey the National Lending Library for Science and Technology found that only a small proportion of 2,355 scientists knew or made use of existing translation services and indexes. A pioneering survey of industrial technologists reported in 1959 that for information on a technical problem only 22 percent of the sample would make straight for the library or some other source of literature.

We have the testimony of Albert Szent-Gyorgyi, Hungarian biochemist and Nobel laureate, to indicate that there is room for improvement here. He writes: 'I could find more knowledge new to me in an hour's time spent in the library than I would find at my workbench in a month or a year'. To set against this we have the oft-quoted warning of Lord Rayleigh, Cambridge physicist and also a Nobel prizeman: 'rediscovery in the library may be a more difficult and uncertain process than the first discovery in the laboratory'. Indeed, Enrico Fermi, Italian nuclear physicist and another Nobel prize-winner, preferred to derive equations himself rather than look them up. He used to wager with his colleagues that he could work them out more quickly than they could find them in the literature.

[2] An excellent account (with bibliography) of such surveys is Margaret Slater ' Meeting the users' needs within the library ' in Jack Burkett *Trends in special librarianship* (Bingley, 1968) 99-136.

THE LIBRARIAN AND THE LITERATURE

Whatever the reasons for this reluctance to make the best use of the literature, whether ignorance, or preference for other channels of communication, or simple disinclination, there is clearly a role here for an intermediary, a guide, and even a teacher. In a paper read to a recent ASLIB Annual Conference, Dr G A Somerfield of the Office for Scientific and Technical Information, a scientist himself, had this to say: ' Perhaps the most important factor restricting the development of better information services is the conservatism of the average user of scientific information and a general lack of understanding of exactly how information is used . . . considerable education of the potential customers is required before they will appreciate the virtues of the different services offered to them '. It is to play at least part of this role that the librarian strives to fit himself.

Of course his approach to the literature does differ from the scientist's or technologist's approach. Only rarely would he attempt to master the primary literature. His stock-in-trade is the secondary literature, in particular the bibliographical and reference tools. As for the tertiary sources, these are peculiarly the librarian's concern, many of them of course having been compiled by librarians for their colleagues. The librarian's particular strength, however, lies in his grasp through systematic study of the underlying pattern discernible in the literature, and his familiarity with the variety in which it manifests itself, combined with an understanding of the relationships of these forms with one another, their comparative reliability, and their varying uses.

THE FORMS OF THE LITERATURE

The forms that the literature has taken and which are listed above are not to be regarded as immutable or eternal. They have no prescriptive right to remain as methods of scientific and technological communication. Even today we can discern overlapping and merging of categories: between dictionaries and encyclopedias, for example, or between yearbooks and directories. And within categories functions and uses change: because of their great increase in size some abstracting services are now found less useful than they used to be for the current approach, although remaining invaluable for the exhaustive approach. It could be argued too that some of the sources listed above as primary should be secondary, and *vice versa*.

New categories appear regularly: citation indexes were unknown to the scientist and technologist ten years ago. New formats too evolve: microfiche, punched cards, magnetic tape. At first these merely record information already available in the regular literature in different form, but the stage is usually reached where the information is *only* available in the new format.

A further hazard in the path of the student striving to understand the structure and forms of the literature is the imprecise way descriptive terms are used in the titles of some scientific and technological publications. J N Friend *Textbook of inorganic chemistry* (Griffin, 1914-37) is a work of twenty four volumes: obviously this is a treatise rather than a textbook in the usually accepted sense of that term. Joan Gomez *Dictionary of symptoms: a medical dictionary* . . . (Centaur, 1967) is not arranged in the alphabetical order that most users expect of a dictionary, but it is set out systematically. In fact, the author of the contributed foreword does comment: 'I feel the term "dictionary" may be misleading'. R G White *Handbook of ultra-violet methods* (New York, Plenum, 1965) is not a handbook in any ordinary sense, but a bibliography. As the introduction says: 'This volume consists of more than 1,600 references . . . arranged alphabetically by senior author'.

In the succeeding chapters the types of scientific and technological literature are described and illustrated. Since this is a textbook the types of literature are discussed in the order in which they should be studied. This is not the evolutionary order outlined above, nor is it necessarily the order the practising scientist would find most useful.

FURTHER READING

S D Beck *The simplicity of science* (Harmondsworth, Penguin Books, 1962).

Ciba Foundation *Communication in science* (Churchill, 1967).

2

GUIDES TO THE LITERATURE

The very first work a student should attempt to locate when commencing his study of the literature of a particular subject is a guide to the literature, if one exists. Fortunately, over the last few years these have greatly increased in number and there are few major disciplines in science and technology that lack some such guide. One that has stood for over forty years and can still stand as a classic example is M G Mellon *Chemical publications : their nature and use* (McGraw-Hill, fourth edition 1965). Guides which attempt to cover the whole field are far less common than those devoted to a single subject area like physics, or atomic energy, or radio-active isotopes. An established major discipline like chemistry may indeed have a dozen such guides.

They are not exactly subject bibliographies in the ordinary sense, because they usually go beyond the normal limits of enumerative bibliography in including not merely lists of references but discussions of the functions and uses of the various types of literature. Some go further: for instance, C R Burman *How to find out in chemistry* (Pergamon, second edition 1966) has a 19-page chapter ' Training and careers '; R T Bottle and H V Wyatt *The use of biological literature* (Butterworths, 1966) has a 22-page chapter ' Libraries and classification '; the guide by Mellon mentioned above has a 78-page chapter on library ' problems ', including 2,500 specific assignments for setting to individual students.

Like many of the tertiary sources, they are surprisingly little known by the scientist or technologist, and it is the librarian or information worker serving the scientist or technologist in whose hands they are most often found. R H Whitford *Physics literature: a reference manual* (Washington, Scarecrow, 1954) is described by its author as ' a borderline contribution between the fields of physics and library science '. Many, indeed, have been compiled by individual librarians:

L J Anthony *Sources of information on atomic energy* (Pergamon, [1966]) draws largely on the author's experience with the United Kingdom Atomic Energy Authority, and is a much expanded version of the pamphlet of the same title first issued by the Atomic Energy Research Establishment Library at Harwell in 1956. Alan Pritchard *A guide to computer literature* (Bingley, 1969) is the outcome of a research project carried out at the North-Western Polytechnic School of Librarianship. Library associations too have produced guides: E B Gibson and E W Tapia *Guide to metallurgical information* ([New York], Special Libraries Association, second edition 1965); and so have the scientists' own professional societies, particularly those with special sections concerned with literature problems: American Chemical Society *Searching the chemical literature* (Washington, 1961).

In some cases (as in the earlier editions of the work by Anthony, just mentioned) it has been the libraries that have produced these guides, primarily for the users of their own collections, but where such collections are substantial the published guides have a value far beyond the walls of a particular library: J E Wild *Patents: a brief guide to the patents collection in the Technical Library* (Manchester Public Libraries, second edition 1966) describes one of the largest collections in Britain outside the Patent Office collection itself. An example from the United States is S Adams and F B Rogers *Guide to Russian medical literature* (Washington, US Public Health Service, 1958) published for the National Library of Medicine.

The majority of these guides are aimed at a wide public—students, teachers, research workers, practising scientists and technologists, as well as information workers and librarians. A number, however, are produced specially for librarians, such as H R Malinowsky *Science and engineering reference sources: a guide for students and librarians* (Rochester, Libraries Unlimited, 1967). Of these, several (such as the present work) are particularly for student librarians, and not uncommonly are for use in particular university or college courses in the literature of science and technology, or are at least based on such courses. F B Jenkins *Science reference sources* (Champaign, Illinois, Illini Union Bookstore, fourth edition 1965), for instance, is 'for use in the course in science reference service at the University of Illinois Graduate School of Library Science', and T P Fleming *Guide to the literature of science* (New York, Columbia University School of Library Service, second edition 1957) is subtitled 'for use in connection with the courses in science literature'.

There are an increasing number of literature courses not for library students but for science and technology students, and we now have examples of guides that have stemmed from these. R J P Carey *Finding and using technical information* (Arnold, 1966) and R C Smith and R H Painter *Guide to the literature of the zoological sciences* (Minneapolis, Burgess, seventh edition 1966) are both in the form of a textbook for just such a course, drawing largely on the teaching experience of the authors, the first in Britain, the second in the United States. R T Bottle *The use of chemical literature* (Butterworths, second edition 1969) is an edited version of actual lectures delivered at Liverpool College of Technology. Practising scientists and technologists too have their courses: Saul Herner *A guide to information tools, methods and resources in science and engineering* (Washington, Herner, 1968) is a recapitulation of the content of a one-and-a-half day course for United States Federal scientists and engineers.

It is probably true to say however that the majority of separately published guides are normal commercial book-trade productions. E J Crane and others *A guide to the literature of chemistry* (Wiley, second edition 1957) claims that its 1927 edition was the 'first comprehensive book to appear in its field'. Another very well known work is N G Parke *Guide to the literature of mathematics and physics* (New York, Dover, second edition 1959). Indeed, with the increase in demand for such guides, particularly from information workers, and the public urgings of the Library Association, ASLIB and the Fédération Internationale de Documentation, a number of publishers have seen the advantage in bringing out such titles in a series. In the United States in 1963 Interscience produced the first of its 'Guides to information sources in science and technology': B M Fry and F E Mohrhardt *A guide to information sources in space and science technology,* a work of close on six hundred pages, even though only current material (1958 to 1962) is included. In Britain, Pergamon's 'International series of monographs in library and information science' includes among its titles guides such as E R Yescombe *Sources of information on the rubber, plastics and allied industries* (1968). Pergamon is also responsible for perhaps the best known series, all entitled *How to find out . . .*, in the 'Commonwealth and international library'.

Not all such guides are separately published in book or pamphlet form. Several hundred have appeared as articles in periodicals, either in the scientific and technical press or in the professional library journals: good examples are W M Hearon 'The literature of cellulose and related materials' *Tappi* 37 September 1954 152A-7A; W I Veasey 'Sources of information in automobile engineering' *Aslib proceedings* 13 1961 167-77; A G Guy 'Sources of metallurgical literature in the Soviet Union' *Special libraries* 51 1960 532-6. Reference books such as encyclopedias and handbooks occasionally contain a literature guide as a separate article or chapter: perhaps the best example is the 50-page article 'Literature of chemical technology' in R E Kirk and D F Othmer *Encyclopedia of chemical technology* (New York, Interscience, 1947-60).

The task of discovering what guides (especially 'hidden' guides of this kind) are available for a particular subject is eased by a comprehensive bibliography which lists well over a thousand, the majority in the form of articles in journals: Gertrude Schutze *Bibliography of guides to the S-T-M literature: scientific-technical-medical* (New York, the author, 1958) and its *Supplements* (1963 and 1967). A more selective listing (itself in the form of a periodical article) is R W Burns Literature resources for the sciences and technologies: a bibliographical guide' *Special libraries* 53 1962 262-71.

The student will already have appreciated that the subjects covered by these individual guides could well range from the whole of science down to the minutest corner of the field, and that in form they might vary from a volume of several hundred pages to a single-page article in a periodical. Obviously, their authors must treat their subjects with varying degrees of intensiveness. Because of the quite different uses to which they are put it is important to learn to distinguish the two main categories of guide:

a) *The 'textbook' type.* Here the emphasis is on exposition, with the stress laid on types of material rather than individual titles: in the preface to B Yates *How to find out about physics* (Pergamon, 1965), for instance, we read that 'no claim is made to comprehensiveness; indeed deliberate policy has been to give indicative selections only'. Another example is G M Dyson *A short guide to chemical literature* (Longmans, second edition 1958) which is deliberately 'non-exhaustive', and where 'no pretence at a complete bibliography is made; the selection of standard works are given only as *examples*'.

b) *The 'reference book' type.* Designed as a working tool, this kind of guide aims at comprehensiveness: the introduction to the Yescombe guide to rubber and plastics information noted above claims that 'every effort has been made to include all important sources'.

Of course examples can be found combining the features of both:

the well-known guide to the literature of chemistry by Crane mentioned earlier states in the preface that it 'is intended to be used both as a reference book and as a textbook. Its coverage of the field of chemical literature is comprehensive . . . The textbook features include discussions of basic principles and topics, emphasis on how to use each form of chemical literature, and an introduction to the art of literature searching.'

A category of guide found exceedingly valuable by the librarian or by any user who is not a subject expert is an elaboration of the 'textbook' type which goes into considerable detail about the *subject* as well as the literature of the subject. Good examples are very few, probably because such works must be very difficult to write, combining as they do an introductory textbook and a guide to its literature, but a model worth studying is Hugo Lemon *How to find out about the wool textile industry* (Pergamon, 1968): chapter 5 of this work, for instance, on 'Wool processing' covers 53 pages but only lists about 150 references to the literature, the bulk of the text being devoted to a summary of the actual processes, with several black and white illustrations.

As with many of the information sources described in the succeeding chapters of this book, and as the last example demonstrates, the boundaries between the types are often diffuse: in particular the line that can be drawn between a guide to the literature and a subject bibliography. In theory the distinction is clear: whatever the arrangement, however detailed the annotations, a bibliography is basically a list of references to the literature; a guide must have something further, as shown in this chapter. In practice this is not so easy: some guides of the comprehensive 'reference book' type are nine parts bibliography, and applying the rule strictly one has to class Walford as a subject bibliography, despite the title *Guide to reference material*.

What is important for the student to be aware of is that this overlapping does exist. He should learn to make his own categorisation of the works he examines, and not to rely entirely on how they are described. He will then not mistake the classic B D Jackson *Guide to the literature of botany* (Index Society, 1881) for a guide to the literature of botany. He will see that it is an uncompromising subject bibliography of about 9,000 unannotated short-title entries.

FURTHER READING

A C Townsend 'Guides to scientific literature' *Journal of documentation* 11 1955 73-8.

3

ENCYCLOPEDIAS

This and the three succeeding chapters are devoted to reference books, and it is worth recalling how R T Bottle, a chemist, has vividly characterised such works: ' One may regard these sources of information as the distilled and highly fractionated final product from the mass of information that has appeared in the preceding periodical literature '. Obviously, these fall into the category of secondary sources, and are all deliberately arranged for ease of consultation. They are the books to which the scientist or technologist turns first for his ' everyday ' information needs (see pages 17-8 above).

Of all reference books the encyclopedia is probably the best known, and the student will already be familiar with the form and function of the great general multi-volumed encyclopedias. Science and technology too have their multi-volumed encyclopedias, with a similar aim: to give in concise and easily accessible form the whole corpus of knowledge within the subject scope of the work. This is an ambitious goal and some would say an impossible task, and encyclopedias have been severely criticised as of little real use to the specialist. Not only does he find his own speciality treated superficially but he knows that by its very form the information is bound to be out of date. This is probably rather unfair, and in any case has less validity if the argument is confined to subject encyclopedias, which can specialise more, and can usually be revised more frequently than the general encyclopedia. Before accepting the criticism it is important for the student to examine with care the various types of encyclopedia in science and technology and to relate their achievements to the demands made on them and the actual use to which they are put.

This so-called ' everyday ' approach to the literature of science and technology is statistically the most frequent: the evidence is unequivocal on this point. It is also an observable fact that the need is commonly for information in a subject area peripheral to the enquirer's primary interest, a field in other words in which he is not a specialist. It is to cater for precisely such demands that many encyclopedias have been compiled, at various levels of subject specialisation. Our largest general science and technology encyclopedia, the 15-volume *McGraw-Hill encyclopedia of science and technology* (second edition, 1966) claims that 'Each article is designed and written so as to be understandable to the non-specialist in its field '. A very different work in single-volume format and in a narrower field, Peter Gray *The encyclopedia of the biological sciences* (Chapman and Hall, 1961), is equally explicit in its aim ' to provide succinct and accurate information for biologists in those fields in which they are not themselves experts '.

To attack encyclopedias for not catering for the specialist is to try to cast them for a role they are by nature unfitted to play. An encyclopedia's task is not to say the last word on a subject, but as has often been said, to give ' first and essential facts ' only. They provide for neither the current nor the exhaustive approach (see pages 16-8), but they do furnish a vast wealth of facts, easily found. They serve the everyday approach, with either specific information or orientation. They may offer no more than a starting point for a closer investigation of a topic, but this too is what they have been designed for, and the bibliographies to the articles can often signpost the way. It should also be borne in mind that reputable scientific and technological encyclopedias are no longer the compilations of hacks: the articles are by experts (2,200 listed in the *McGraw-Hill encyclopedia* and 500 in Gray), and are usually signed by them.

ENCYCLOPEDIAS FOR THE LAYMAN

It should not be too readily assumed, however, that encyclopedias of this kind are for the layman: neither of these two examples are. A number of the articles run to several pages, and into much technical detail. The preface to the *McGraw-Hill encyclopedia* is specific about level: ' Most of the articles, and at least the introductory part of them, are within the comprehension of the college undergraduate in science or engineering '.[1] That is not to say that the intelligent non-scientist could not get anything out of this work, but it does remind us of one very important role of the encyclopedia in all fields,

[1] It is interesting to note that in the second edition this reads: ' within the comprehension of the interested high school student '.

namely to explain its subject to the 'ordinary enquirer', for it is often to an encyclopedia of the subject that such a man would turn first. Moreover, in the field that concerns us here this is particularly vital, for science and technology are critical in shaping our world. This is the sphere in which the encyclopedia editor is seen, in the words of Lowell A Martin, American librarian and educator, as 'a mediator, between the world of scholars on the one side and the individual seeking information on the other, between those who know something and those who seek to know'.

And indeed, in science and technology, encyclopedia editors have taken up this challenge, and we are seeing increasing numbers of high-quality examples of this second category of encyclopedia, for the non-scientist. The editor's introduction to J R Newman *The international encyclopaedia of science* (Nelson, 1963) says firmly that 'the needs of the common reader—the student, the teacher, the non-specialist—have been our measuring rod'. This work can also serve as an object lesson to those who may believe that a scientific encyclopedia for the non-scientist must be inferior in some way. In fact there is no intrinsic reason why such a work should not maintain the same standards of scholarship and reliability as the best of similar works for a more specialist readership. This work by Newman, indeed, demonstrates its standards by having its articles signed by a team of 450 scientists and engineers of some position. And it is likely that, if asked, they would confirm a common experience of scientific writers: it is often far more difficult to write for the layman than for fellow-scientists.

An interesting example of a single-volume encyclopedia for the broad field of science and technology, written by a scientist of repute, is J G Cook *Science for everyman encyclopaedia* (Watford, Merrow, 1964): this is aimed at the home library, and in particular the student (there is a 72-page 'Student's reference section' at the end, comprising mainly tables). Many of the layman's encyclopedias, however, are devoted to more specialised subjects, which although indisputably scientific or technological, have become traditionally the province also of the amateur enthusiast: *The encyclopaedia of radio and television* (Odhams, second edition 1957), although quite technical, is for the amateur; similarly the preface of the two-volumed R Sudell *Odhams garden encyclopaedia illustrated* (1961) tells us of its aim 'to help the amateur who would like to know more about the plants he grows and the principles that underlie their cultivation'; and clearly E F

Carter *The railway encyclopedia* (Starke, 1963) is for the lay enthusiast rather than the professional railwayman.

ENCYCLOPEDIAS FOR THE SPECIALIST

All this is not to imply that there are no encyclopedias for the specialist. In point of fact, as the subject field narrows it obviously becomes possible to cater more for the experts, and in numbers this category of specialist subject encyclopedia is probably the largest of the three examined in this chapter. The student who examines the massive *Kirk-Othmer encyclopedia of chemical technology* (Interscience, second edition 1963-), in 18 volumes, can be left in no doubt that this is a tool for professional chemists and chemical engineers. Indeed, it is clear to see in this and similar comprehensive works an extension of the usual role of the encyclopedia beyond the 'first and essential facts' only: although still alphabetically arranged, they share to some extent the nature of a treatise, that is to say an authoritative attempt to digest all the primary literature and to consolidate the whole of existing knowledge on a topic. From the same publisher as *Kirk-Othmer* we have the very similar *Encyclopedia of polymer science and technology: plastics, resins, rubbers, fibers* (New York, 1964-), to be in 10 or 12 volumes, and *Encyclopedia of industrial chemical analysis* (1966-), of about the same size. The editors of both see their subjects as ripe for this kind of thorough-going approach: in the words of the preface to the second work, 'A comprehensive encyclopedic treatment of the present state of the art seems to be a desirable and worthwhile undertaking'. Of course the physical form they take inevitably means that they cannot furnish absolutely current data: their function is to act as the great repositories of received knowledge. In fact, no sooner was the first edition of *Kirk-Othmer* complete (1947-60 in 15 volumes and 2 supplements) than plans were under way for the second, and we have already been warned that this will be necessary also for the *Encyclopedia of polymer science and technology*.

These huge syntheses of a complete area of knowledge, however, are not typical of the usual special subject encyclopedia. The majority are handy single-volume works, representing not only a remarkable range of subjects but an astonishing variety of content. Typical examples are M G Say *Newnes' concise encyclopaedia of electrical engineering* (1962) with just under a thousand pages; *The universal encyclopedia of mathematics* (Allen and Unwin, 1964) with extensive sections of formulae and tables; N W Kay *The modern building encyclopaedia* (Odhams, 1955), a work of a very practical nature, notable for its illustrations, averaging two for each of its 768 pages. A number of publishers have seen the advantages of bringing out such encyclopedias in a series, and the most extensive range at present is the couple of dozen Reinhold/Chapman and Hall volumes now so familiar in our libraries in their handsome standard format. They do vary to some extent one from another, but in general they contain longish signed articles (perhaps 400 in a 1,200-page volume) by experts, with short, selective bibliographies, and are deliberately aimed at a wide spectrum of users, from school students (in the higher forms), and teachers, to specialists. Examples to examine are R J Williams and E M Lansford *The encyclopedia of biochemistry* (1967), C A Hampel *The encyclopedia of the chemical elements* (1968), or H R Clauser *The encyclopedia of engineering materials and processes* (1963). Other volumes are devoted to, for example, microscopy, x-rays, oceanography, chemical process equipment; the one on chemistry by G L Clark and G G Hawley has reached a second edition (1966).

Indeed the non-librarian is often staggered to discover some of the specialised topics in which there exists a fully-fledged encyclopedia: E Gurr *Encyclopaedia of microscopic stains* (Hill, 1960), R H Durham *Encyclopedia of medical syndromes* (Harper and Row, 1966), W M Levi *Encyclopedia of pigeon breeds* (Jersey City, TFH Publications, 1965) are three examples. It must also be recorded that there are a number of important disciplines still lacking an encyclopedia (at least in English): anthropology and computers are two which come to mind at the time of writing.

CONTENT OF ENCYCLOPEDIAS
It is well worth the student's while paying some attention to the various kinds of information contained in encyclopedias, for they are by no means uniform in content. As well as the ' short article ' and ' long article ' approaches with which he will be familiar from his study of general encyclopedias he will find for example that it is not uncommon for editors to include substantial amounts of data outside the basic alphabetical sequence of the text: A K Osborne *An encyclopaedia of the iron and steel industry* (Technical Press, second edition 1967), for instance, has 48 pages of appendices with conversion tables, signs and symbols, and information about societies in the subject field.

The number and position of bibliographical references also varies quite markedly. The special-subject single-volume works for the lay enthusiast hardly ever indicate further sources of information, and even the more substantial works covering the whole field of science and technology for the non-expert rarely list any references. The remarkable *Van Nostrand's scientific encyclopedia* (fourth edition 1968) packs nearly two-and-a-half million words and over two thousand pages between the covers of a single volume, but has no bibliographies. The Newman *International encyclopedia* referred to earlier lacks bibliographies with the four thousand individual articles, but does have a 13-page 'graded reading list' in volume four. Most of the better encyclopedias for the specialist, on the other hand, carry bibliographies with the articles as a matter of course: with works of *Kirk-Othmer* calibre these are often as extensive as any list of references appended to a monograph or a paper in a learned periodical. Yet another approach is seen with the *McGraw-Hill encyclopedia:* even though most of the longer articles have their expected bibliographies there is also the separately published *McGraw-Hill basic bibliography of science and technology,* designed as a supplement to and uniform with the volumes of the main set. Books are listed under alphabetically-arranged subject headings corresponding to entries in the *Encyclopedia.*

Further study would illustrate varying approaches among encyclopedia editors to such topics as biographical information, historical aspects of the subject, illustrations, etc.

FORM OF ENCYCLOPEDIAS

Library users are constantly surprised to come across encyclopedias that are not alphabetically arranged, such as D R Woodley *Encyclopaedia of materials handling* (Pergamon, 1964) in two volumes, or the *Universal encyclopedia of machines; or how things work* (Allen and Unwin, 1967). And yet historically the systematically arranged encyclopedia was the first on the scene by many hundreds of years. The justification for this arrangement is succinctly given in the preface to the ambitious new *Materials and technology: a systematic encyclopaedia* (Longmans, 1968-), where the claim is made that it 'ensures that related subjects are dealt with in proximity with each other rather then separated by the random vagaries of the alphabet '. Scheduled for completion in eight volumes by 1972 this will be the largest systematic encyclopedia of science and technology in English and it will be interesting to observe its use, particularly as a further claim is to furnish ' up-to-date information for the layman and technologist alike '. It has been ' written specifically with the intention that each subject should be capable of being fully understood by a person who is not an expert in that subject '. As a source of everyday reference the key to success for a systematically arranged work lies in its index, and *Materials and technology* is being published with an index in each volume. It is for this reason that S S Kutateladze and V M Borishanskii *A concise encyclopedia of heat transfer* (Oxford, Pergamon, 1966) is so frustrating for the user, for not only are the contents not in alphabetical order, but there is no index to its more than five hundred pages!

What is more disturbing about this example, however, is that it is not really an encyclopedia at all, systematic or otherwise. The current Pergamon catalogue describes it as a ' collection of data and formulae used in calculations on all types of heat transfer problems '. A more accurate description therefore would seem to be ' handbook ', and this is borne out by the wording of the original Russian title which is best translated as ' reference book '. The student should be on his guard for the occasional examples of this imprecision in titles. None of the following, for instance, are encyclopedias: T Corkhill *A concise building encyclopaedia* (Pitman, 1951) is a dictionary of about 14,000 terms; the *International encyclopedia of physical chemistry and chemical physics* (Pergamon, 1960-) is the collective title of a series of monographs such as B Donovan *Elementary theory of metals* (1967) and E R Lapwood *Ordinary differential equations* (1968); the ' *Modern plastics* ' *encyclopedia* (New York, Plastic Catalogue Corp) is a yearbook and directory issued as a supplement to the American technical journal *Modern plastics*.

Conversely there are encyclopedias to be found masquerading under other titles. J F Hogerton *The atomic energy deskbook* (Reinhold, 1963) is an encyclopedia by any definition, and is, moreover, alphabetically arranged; the author's preface describes it as ' combining the features of a dictionary and an encyclopedia '. A L Howard *A manual of the timbers of the world* (Macmillan, third edition 1948) is a similar case. But by far the most commonly confused ascription is ' dictionary '. Dozens of examples could be quoted of works described as dictionaries that are really encyclopedias, but two will suffice : *Thorpe's dictionary of applied chemistry* (Longmans, 1937-56) in eleven amply referenced volumes plus an index volume was the predecessor of *Kirk-Othmer* in its field; and S K Runcorn *Inter-*

national dictionary of geophysics (Pergamon, 1967) comprises two large volumes of 'concise authoritative articles' (*eg*, 38 pages on 'earthquakes') by specialists, arranged alphabetically with bibliographical references.

The student will know that in theory the distinction between an encyclopedia and a dictionary is quite clear: it is demonstrated quite neatly in the twin works by Peter Gray. In *The encyclopedia of the biological sciences* already quoted (page 28 above) he writes: 'This is an encyclopedia, not a dictionary. That is, it does not merely define the numerous subjects covered but describes and explains them'. Five years later in *The dictionary of the biological sciences* (Reinhold, 1966) he explains: 'It was the infeasibility of indexing the "Encyclopedia of the biological sciences" in a manner that would permit enough individual words to be found that led me to the conviction that a separate dictionary was a necessity'.

In practice this precise line is very hazy. We have companion volumes from the same publisher, identical in format and virtually identical in form, but one is *The international dictionary of physics and electronics* (Van Nostrand, second edition 1961) while the other is *The international encyclopedia of chemical science* (Van Nostrand, 1964). There are works which try to have it both ways: the Royal Horticultural Society *Dictionary of gardening* (Oxford, Clarendon Press, second edition 1956) in four volumes and its *Supplement* (1956) is subtitled 'a practical and scientific encyclopaedia of horticulture'. And there are examples that are genuinely both: E J Labarre *Dictionary and encyclopaedia of paper-making* (OUP, second edition 1952) combines a dictionary (in English with equivalents in six other languages) with an encyclopedia (the article on 'wallpaper', for instance, comprises eleven pages and nine plates).

What has also emerged is a cross-bred article known as an encyclopedic dictionary, although it only needs two examples to demonstrate the difficulty of achieving a satisfactory definition: R I Sarbacher *Encyclopedic dictionary of electronics and nuclear engineering* (Pitman, [1960]) defines 14,000 terms within one volume; J Thewlis *Encyclopaedic dictionary of physics* (Pergamon, 1961-4) is a nine-volume work, 'of graduate or near-graduate standard', aiming to 'put the whole of physical knowledge on the bookshelf'. It is not easy to see how both these 'encyclopedic dictionaries' could be encompassed in one category of reference book.

The difficulty of ensuring that the information in an encyclopedia is as up-to-date as possible has been touched on more than once in this chapter: with so substantial and permanent a work as the last example the size of the problem can be imagined. In fact the publishers have gone about this in a straightforward and workmanlike fashion, and each year since the completion of the main set have issued a supplementary volume of over four hundred pages in the same form as the original ' Thewlis ', each complete with its own index. It is planned to cumulate these supplementary indexes every five years.

The alternative to frequent supplements is of course a new edition. Naturally this is easier for both editorial and production reasons where the encyclopedia is a small or single-volume work: *Kingzett's chemical encyclopaedia* (Bailliere, ninth edition 1966) has relied on this method since its first publication in 1919.[2]

In theory the most efficient method but in practice the most cumbersome is the loose-leaf encyclopedia: new information can be added in any quantity at any time at any point in the work and superseded information can be likewise discarded. Examples are rare, but one that will repay study is the *Faraday encyclopedia [of] hydrocarbon compounds* (Manchester, Chem-index, 1949-).

---

[2] It is worth noting that both *Kirk-Othmer* and *McGraw-Hill* seem to be combining both methods: supplements (called ' year books ' by *McGraw-Hill*) together with a new edition shortly after the first.

4

DICTIONARIES

As one of our most common reference books, the dictionary is probably less in need of explanation than any other. Its concern is words: either the general words of a language, or, as in this case, the special terms of a particular subject discipline. In a field like science and technology, so dependent by its very nature on communication, there is no need to stress the crucial importance of words. What is worth recalling, however, is that increasingly over the last few generations, scientists and technologists have found the ordinary language of scholarly converse inadequate to convey what they have to say to one another. We have now arrived at a point where, to quote the preface to *Chambers's technical dictionary* (third edition 1958), ' To be safe, indeed, one must regard technical language as a language apart from ordinary speech. Technical terms are in reality symbols adopted, adapted, or invented by specialists and technicians to facilitate the precise expression and recording of their ideas '.

The extent of this ' language apart ' can perhaps be gauged by the fact that there are now several thousand dictionaries in the field, and substantial volumes have been compiled devoted solely to the vocabulary of very specialised topics: Raymond Jahn *Tobacco dictionary* (New York, Philosophical Library, 1954) and E Bruton *Dictionary of clocks and watches* (Arco, 1962) both have over two hundred pages; L L Copeland *The diamond dictionary* (Los Angeles, Gemmological Institute of America, 1960) and A E Dodd *Dictionary of ceramics* (Newnes, 1964) have over three hundred. Furthermore, since the first edition of *Chambers's* in 1940 with its 60,000 terms there has been no serious attempt at a comprehensive dictionary of the whole field of science and technology. Some hold the view that there is never likely to be such a work again, so vast and rapidly changing is the vocabulary. Such as are currently available are highly selective; F S Crispin *Dictionary of technical terms* (Milwaukee, Bruce, tenth edition 1964), for instance, contains only about ten thousand terms mainly from the applied sciences. Indeed in this general field of science and technology the majority of titles are aimed at the non-expert, either the layman or the student. E B Uvarov and D R Chapman *A dictionary of science* (Harmondsworth, Penguin, third edition 1966) is one of an extensive series of Penguin dictionaries for the non-specialist, with about five thousand words. A Hechtinger *Chatto's modern science dictionary* (1961) is a more substantial work of over five hundred pages, compiled in the United States for school students, and defining ' in uncomplicated language the meanings of some 16,000 scientific and technical words '. W E Flood and Michael West *An elementary scientific and technical dictionary* (Longmans, third edition 1962) is a remarkable compilation, defining some ten thousand words using an explaining vocabulary confined to two thousand words (only sixty of which are technical). The aim is to avoid giving definitions which require a dictionary to be understood! Special attention is given to word-elements (roots, suffixes, prefixes) ' so that the reader may be enabled to break up and interpret new scientific terms for himself '. A rather special example using an even smaller vocabulary to define 25,000 terms is E C Graham *The science dictionary in Basic English* (Evans, 1965).

It is to dictionaries in his speciality, however, that the practising scientist or technologist turns more readily, and it is these that make up the bulk of titles in descriptive lists such as W R Turnbull *Scientific and technical dictionaries: an annotated bibliography* (San Bernardino, Calif, Bibliothek Press, 1966-). Classic works of this kind well worthy of study are *Hackh's chemical dictionary* (New York, McGraw-Hill, fourth edition in preparation), with 80,000 entries; I F and W D Henderson *A dictionary of biological terms* (Oliver and Boyd, eighth edition 1963) first published in 1920; J G Horner *A dictionary of mechanical engineering terms* (Technical Press, ninth edition 1967).

It is common for such dictionaries to go beyond their basic function of defining words: Glenn and R C James *Mathematics dictionary* (Van Nostrand, third edition 1968) claims to be ' by no means a mere word dictionary . . . It is rather a correlated condensation of mathematical concepts, designed for time-saving reference work '. The most usual way is for the compiler to extend the individual entries under each word to include further information on the subject, but a more radical method is to include extra entries, quite overtly encyclopedic in kind: L M Miall and D W A Sharp *A new dictionary of chemistry* (Longmans, fourth edition 1968) and H J Gray *Dictionary of physics* (Longmans, 1958) are companion volumes with articles sometimes extending to a page or more, including a number of biographical entries, and occasional literature references. And examples abound of dictionaries with substantial appendices of tables, formulae, and other non-dictionary data: a quarter of the four hundred or so pages of *Chambers's dictionary of electronics* (1969) is taken up thus.

Taking this into account with the evidence in the previous chapter on the imprecise way in which the term ' dictionary' is applied to works transparently encyclopedic, the student must be aware that in certain subjects for most practical purposes the distinction between encyclopedia and dictionary no longer has any meaning. The terms are often used interchangeably, sometimes within the same work, as in J R Stewart and F C Spicer *An encyclopedia of the chemical process industries* (New York, Chemical Publishing Co, 1956) which is the latest version of what used to be called *Stewart's scientific dictionary:* and likewise in the *International encyclopedia of chemical science* (Van Nostrand, 1964) which replaces *Van Nostrand chemist's dictionary* (1954). This is not to suggest, however, that the student should not cultivate an appreciation of the special role of the dictionary in scientific and technological communication. In any field in which he wishes to specialise he must be able to distinguish those ' dictionaries ', however they may be described, which are in fact concerned with terminology from those that are mainly encyclopedias.

Among the many problems facing the compiler of a subject dictionary there are two in particular that it is important to have in mind when examining an individual title. The first is the question as to what words to include. To some the answer might seem obvious: those words on the one hand that lie outside the general vocabulary and do not appear in an ordinary dictionary, such as ' thermistor ', ' bus-bar ', ' thyratron ', and on the other hand those words in common usage that have acquired a specialised meaning within a particular field, such as ' resistance ', ' valve ', ' terminal '. For example, the preface to Michael Abercrombie *A dictionary of biology* (Harmondsworth, Penguin, fifth edition 1966) states that ' semi-technical terms, which can be found in any English Dictionary . . . are omitted '. Yet it can be seen that some compilers do include ordinary words in their ordinary meanings: Walford points out that E J Gentle and C E Chapel *Aviation and space dictionary* (Los Angeles, Aero Publishers, fourth edition 1961) contains definitions of words like ' electrician' and 'telephone'; and *Audel's New mechanical dictionary* (New York, 1960) defines 'sizable' and 'dry'. As for including words from related disciplines that users of a dictionary might find useful, there are two points of view: some workers prefer the subject scope of their dictionaries clear-cut; they wonder whether terms from other fields can be defined with quite the same authority; and they know that they can only be chosen on a very selective basis. But many users do appreciate this extension of coverage, particularly as boundaries between subjects grow ever more indistinct, and many dictionary compilers do follow the practice, as for example George Usher *A dictionary of botany* (Constable, 1966) which is subtitled 'including terms used in biochemistry, soil science and statistics'.

The second problem of a dictionary compiler that the librarian in particular needs to consider has to do with the level of readership at which the book is aimed, for this obviously has considerable bearing on the kind and detail of terminology used in definitions and entries. This can be illustrated by taking a well-established discipline such as mathematics and studying the range of dictionaries that are available. At the simplest level is C H McDowell *A dictionary of mathematics* (Cassell, 1961) 'written specially for young people'. For students at the secondary school level we have J Bendick and M Levin *Mathematics illustrated dictionary* (New York, McGraw-Hill, 1965). A work like W Karush *The Crescent dictionary of mathematics* (Macmillan, 1962) covers high-school and college mathematics in detail. According to its preface C C T Baker *Dictionary of mathematics* (Newnes, 1961) is 'suitable for use up to degree standard', while W F Freiburger *The international dictionary of applied mathematics* (Van Nostrand, 1960) is for the advanced scientist.

Aside from questions of scope and level, it would be a mistake to assume that all dictionaries in science and technology are fairly similar. There may not be the variety of types that one finds among the general language dictionaries, such as dictionaries of slang, spelling dictionaries, rhyming dictionaries, and the like, but the range is surprisingly wide. In W E Flood *The origins of chemical names* (Oldbourne, 1963) entitled *The dictionary of chemical names* in the United States, we have an etymological dictionary, giving derivations for the words listed. John Challinor *Dictionary of geology* (Cardiff, University of Wales, third edition 1967) is, like the *Oxford English dictionary*, compiled 'on historical principles', with quotations showing each term in a variety of uses. E C Jaeger *The biologist's handbook of pronunciations* (Springfield, Ill, Thomas, 1960) concentrates on pronunciation to the virtual exclusion of definitions.

Featured in many of the titles so far quoted, and particularly useful in the field of science and technology, are illustrations, usually simple black-and-white diagrams or line-drawings, *eg*, Petroleum Educational Institute *Illustrated petroleum dictionary* (Los Angeles, 1952). Especially valuable are those dictionaries based on more elaborate drawings, often showing ' exploded ' views, with each individual part labelled and keyed to the text, *eg*, *Handy technical dictionary in 8 languages* (KLR Publishers, 1949).

Dictionaries of synonyms are apparently not known in science and technology, but in recent years there have appeared a number of thesauri, which in many ways resemble synonym dictionaries, for example, *The thesaurus of engineering and scientific terms* (New York, Engineers Joint Council, 1968), and the National Agricultural Library *Agricultural/biological vocabulary* (1967) in two volumes with *Supplement* (1968). Of course these are not produced *as* dictionaries but as subject-heading lists for indexers and searchers in particular information retrieval systems, both manual and automated. Indeed, the American Society for Metals *Thesaurus of metallurgical terms* (Metals Park, Ohio, 1968) was first available in computer-printout form only. The value of the thesaurus arrangement, however, has been seen by at least one compiler of a conventional dictionary: Peter Gray *The dictionary of the biological sciences* (Reinhold, 1966) makes extensive use of the practice of grouping words according to their meaning rather than alphabetically.

A problem for users of language in all fields is that of unifying and standardising terms and their definitions. France has her Académie Française and Italy her Accademia della Crusca to care for the purity of their national tongues, but English has never had such a body. Yet in science and technology the need has become so acute that a number of national and international organisations have taken it upon themselves to bring some order into the chaos of scientific and technical terminology. In many cases their efforts have taken the form of an official list of standardised and agreed terms, virtually an ' official ' dictionary. An outstanding example is the American Geological Institute *Glossary of geology and related sciences* (Washington, second edition 1960) and *Supplement* (1961): close on twenty thousand terms are authoritatively and officially defined; where the ' Committee believes use should be abandoned ' the entries are marked ' Not recommended '. The British Standards Institution, working through its numerous committees, is very active in this sphere, and has produced over a hundred glossaries: for example, *Glossary of terms relating to air-cushion vehicles* (BS 4236:1967) which defines 42 terms; *Glossary of terms used in the rubber industry* (BS 3558:1962); *Typeface nomenclature and classification* (BS 2961:1957); *Definitions for use in mechanical engineering* (BS 2517:1954), and (in several parts) *Welding terms and symbols* (BS 499: 1965). Two interesting examples of official dictionaries are United States Air University *The United States Air Force dictionary* (Princeton, NJ, Van Nostrand, 1956) and Institute of Petroleum *A glossary of petroleum terms* (third edition 1961). A handy guide is the UNESCO *Bibliography of monolingual scientific and technical glossaries* (Paris, 1955-9): the first of its two volumes lists those titles that have been approved by the various national standards associations.

As the student will notice, the term that has been used most commonly in the titles of these works is ' glossary '. Of course the word is far older even than ' dictionary ', and means no more in fact than a list, with explanations, of specialised terms, but it is a convenience to have it applied to this particular kind of standardising dictionary. The student should be warned, however, that this description, like so many others, is not used consistently: Saul Patai *Glossary of organic physical chemistry* (Interscience, 1962), the *Meteorological glossary* (HMSO, fourth edition 1963), A W Lewis *A glossary of woodworking terms* (Blackie, 1966), T Corkhill *A glossary of wood* (Nema, 1948), and very many others, are no more than ordinary dictionaries.

There is not the space here to go into the related but vast and thorny problem of nomenclature which plagues certain disciplines like chemistry and botany and zoology. In Britain, for example there are over 70 names for the heron. Names for new chemical compounds are a constant headache for research chemists, for with new chemical compounds being synthesised in their thousands it is obviously important that they should be named as systematically and descriptively as possible. Unfortunately, no agreement has yet been reached by the chemists on an international basis of nomenclature: an index prepared by the Chemical Abstracts Service showed that 33,000 chemical substances were described in the literature by 140,000 names. What the student should know is that efforts are being made at systematisation, exemplified by publications such as the International Commission on Zoological Nomenclature *International code of zoological nomencla-*

ture (1961) which 'indicates the criteria governing the naming of an animal or group of animals, and also regulates the use of names which have already appeared in the literature', and the British Standards Institution *Recommendations for the selection, formation and definition of technical terms* (BS 3669 : 1963).

FORM OF DICTIONARIES

When asking in a library for a dictionary of a particular subject most readers expect to be given a book, usually a single bound volume, devoted to their topic. In fact there are hundreds of dictionaries published as part of another work, either a reference book or a book 'in the field'; and probably almost as many not in book form at all. In A K Graham and H L Pinkerton *Electroplating engineering handbook* (New York, Reinhold, second edition 1962) there is a 9-page glossary, and there is a 24-page list of technical terms in A W Eley *Stockings, silk, rayon, cotton and nylon* (Leicester, Hosiery Trade Journal, 1953). D L Bloem 'Glossary of terms on cement and concrete technology' appeared in instalments in the *Journal of the American Concrete Institute,* from December 1962, and T Armstrong and B Roberts 'Illustrated ice glossary' was first published as two articles in *Polar record* for January 1956 and May 1958.

' STRUCTURAL ' DICTIONARIES

As virtually a separate language within the mother tongue, scientific and technological English has been the object of the same scholarly attention that other variations of standard English have attracted, and we now have a number of books and papers like T H Savory *The language of science* (Deutsch, 1953) and P B McDonald 'Scientific terms in American speech' *American speech* 2 November 1926 67-70. We even have works of instruction like R W Brown *Composition of scientific words: a manual of methods and a lexicon of materials for the practice of logotechnics* ([Baltimore], the author, 1954): 'logotechnics' the author defines as 'the art of composing words' and his aim is to 'diminish the area and amount of verbicultural wrongdoing'. The bulk of the volume of close on nine hundred pages is taken up with an etymological dictionary.

A more common linguistic approach is that already described above in Flood and West *An elementary scientific and technical dictionary* (page 37) in which special attention is given to the elements that go to make scientific and technological words. Another dictionary by

W E Flood, *Scientific words, their structure and meaning* (Old-bourne, 1960), is devoted entirely to 'about 1,150 word-elements (roots, prefixes, suffixes) which enter into the formation of scientific terms'. The entry under ' stetho-' derives this from the Greek ' stethos ' meaning ' the chest '; the entry for ' scope ' traces this to the Greek ' skopos ', meaning ' one who watches '. A stethoscope therefore is an ' instrument for " inspecting the chest " '. There are several examples of these ' structural ' dictionaries in particular subject areas, for example, E C Jaeger *A source-book of biological terms* (Springfield, Ill, Thomas, third edition 1955) lists 15,000 of these linguistic building bricks.

Much terminology, old and new, is of course based on Greek and Latin elements, particularly in fields like medicine and botany, yet it is common to find that many workers have, like Shakespeare, ' small Latin and less Greek '. Special attention is given to this in dictionaries like R S Woods *The naturalist's lexicon: a list of classical Greek and Latin words used or suitable for use in biological nomenclature* (Pasadena, Abbey Garden Press, 1944) and *Addenda* (1947). Further assistance is available from texts like W T Stearn *Botanical Latin: history, grammar, syntax, terminology and vocabulary* (Nelson, 1966) which claims that the language is ' now so distinct from classical Latin in spirit and structure as to require independent treatment '.

TRANSLATING DICTIONARIES

One tremendous problem arising directly from the demise of Latin as the language of science (and it should not be forgotten that as late as 1687 Newton used Latin for his greatest work, *Philosophiae naturalis principia mathematica*) is the ' language barrier ', which will be discussed at length in chapter 16. Our immediate concern is with those thousands of interlingual dictionaries, compiled as aids to translation. There are fortunately a number of lists to help the librarian thread his way through the mass of titles : perhaps the most informative of several available is the UNESCO *Bibliography of interlingual scientific and technical dictionaries* (Paris, fourth edition 1961) and *Supplement* (1965), although lists of an individual library's own holdings can be very useful on the spot, particularly if carefully indexed. A good example is Manchester Public Libraries *Technical translating dictionaries* (1962) which includes almost four hundred titles in alphabetical subject order with a language index.

Much of what has been said earlier in this chapter about monolingual dictionaries applies with equal force to translating dictionaries, such as the difficulty of compiling a really satisfactory work covering the whole field, although there have been two substantial attempts recently by major publishers with A F Dorian *Dictionary of science and technology, English-German* (Elsevier, 1967) and Rudolf Walther *Polytechnical dictionary, English-German* (Pergamon, 1968) and *German-English* (1968). Publicity for the latter does bring home just how much is required of such a dictionary, for it makes the ambitious claim that it ' should bridge the gap between the general, non-technical and the specialised subject dictionaries. Terms from all fields of science and technology are included, and the selection provides both a basic vocabulary common to many fields supplemented with vocabularies peculiar to those technological sciences which form the basis of several specialized fields. Some relevant general terminology has also been included '.

In practice, however, specialist users prefer a dictionary of narrower subject scope, preferably one confined to their speciality. Highly re-garded and widely used titles are the two-volumed *Castilla's Spanish and English technical dictionary* (Routledge, 1958) which excludes the physical, chemical and biological sciences; Louis De Vries and T M Herrmann *German-English technical and engineering dictionary* (McGraw-Hill, second edition 1965) and *English-German technical and engineering dictionary* (McGraw-Hill, second edition 1968), both of which exclude scientific terms; and the even more specialised Richard Ernst and E I Morgenstern *Dictionary of chemistry: German-English* (Pitman, 1961) and *English-German* (1963); and United States Department of the Army *English-Russian, Russian-English electronics dictionary* (New York, McGraw-Hill, 1958).

Perhaps because the general bilingual dictionaries used by language students and scholars are usually two-way (*eg*, from English into French *and* from French into English), it is commonly found (and the student will already have noticed) that many scientific and technical dictionaries are two-way also. Yet there is much evidence that use is largely one-way: the English-speaking chemist will almost invariably wish to translate papers and books *from*, say German or Russian. The professional translator likewise will spend most of his time turning material *into* his own language, whether French or Chinese or any other. Some of the finest translating dictionaries are indeed one-way only: the student should study with care L I Callaham *Russian-English chemical and polytechnical dictionary* (Wiley, second edition

1962); A M Patterson *A German-English dictionary for chemists* (New York, Wiley, third edition 1950); and the companion volume A M Patterson *A French-English dictionary for chemists* (New York, Wiley, second edition 1954).

Representatives of most of the categories of dictionary described earlier can also be found among the translating dictionaries. Indeed, the illustrated dictionary using diagrams with keyed parts was pioneered by the monumental Schlomann-Oldenbourg series *Illustrated technical dictionaries in six languages* from various publishers since 1906. There are many examples of interlingual glossaries of standard terms: the International Commission on Glass *Dictionary of glass-making* (Charleroi, 1965) is in English with French and German equivalents, and indicates incorrect terms by inverted commas. English, French and Russian are the language of the International Organization for Standardization *List of equivalent terms used in the plastics industry* ([Geneva], 1961). And reminding us that today it is not only man that communicates with his fellows we have Hans Breuer *Dictionary for computer languages* (Academic Press, 1966): the user is told that ' using this book as one would a common translation dictionary, it is possible to translate a FORTRAN program into ALGOL, and vice versa '.

The observant reader will doubtless have noted that some of the titles just quoted are multilingual (or polyglot) dictionaries. It is instructive to consider the organisational problems of a dictionary compiler once his field extends to more than two languages, and to examine some of the proffered solutions. The method adopted by the famous *Hoyer-Kreuter technological dictionary* (Berlin, Springer, sixth edition 1932-4) is simple but expensive: a separate and self-contained edition for each of the three languages, German, English, and French, each edition of course giving equivalents in the other two languages. This approach is quite unrealistic, however, where the dictionary attempts to cover five or six or seven languages, as is frequently the case. The system used by the very successful Elsevier multilingual series is to arrange the volume by the English word, followed in tabular form by the definition, and then by the equivalent word in the other five or six languages in turn. Individual indexes for each of these languages refer (usually by number) to the English word equivalent. Well-established examples to study are W E Clason *Elsevier's dictionary of television, radar and antennas in six languages* (1955), and G S Stekhoven *Elsevier's automobile dictionary in eight languages* (1960). Several in this series have been extended from, say, six languages to seven by a supplement, *eg*, W E Clason *Elsevier's dictionary of nuclear science and technology in six languages* (1958) has a *Russian supplement* (1961). Other variations include classified rather than alphabetical order in *Elsevier's dictionary of the gas industry in seven languages* (1961); and the use of German rather than English as the base language, as in B D Hartong *Elsevier's dictionary of barley, malting and brewing in six languages* (1961). Elsevier were not of course the inventors of this method, nor are they the only publishers to adopt it: J Nijdam *Horticultural dictionary in eight languages* (Interscience, [third edition] 1962) is compiled on a Dutch base, and was first published by the Netherlands Ministry of Agriculture in 1955.

Dictionaries of this kind have come in for two specific criticisms: they include only a comparatively small number of terms (for the tabular layout consumes a great deal of space), and the ' synoptic ' coverage oversimplifies the translation problem (for the validity of a ' neat row of five foreign equivalents ' is suspect). Nevertheless, they do seem to fulfil a need, and they proliferate greatly, even in quite specialised subject fields, for instance, H A Anderfelt *Technical vocabulary for the match industry: English, Swedish, German, French.* (Jonkoping, Swedish Match Co, 1961), and J Brandt *Emails, enamels, émaux, smalti: a dictionary in four languages* (Leverkusen, Farbenfabriken Bayer Ag, 1960). There are even works which give equivalents in a dozen or more languages: G Carrière *Detergents: a glossary of terms* (Van Nostrand, 1960) lists only 257 terms, but covers 19 languages; A Herzka *Elsevier's lexikon of pressurized packaging* (*aerosols*) (1964) lists 262 terms in 21 languages!

It should not be too complacently assumed that English, even scientific and technological English, means the same thing throughout the world. There are considerable variations between British and American English, for instance: a car hood is two very different things on either side of the Atlantic; many drivers in Britain would be puzzled to learn that American cars use gas for fuel; even a gallon is not the same. It is perhaps not yet necessary to have a British-American translating dictionary, but many of the scientific and technological dictionaries do take trouble to distinguish the two usages. It is standard practice in the Elsevier multilingual series, and is given particularly close attention in Magda Polanyi *Technical and trade dictionary of textile terms* (Pergamon, second edition 1967), German-English and English-German.

A final note of caution on translating dictionaries: only rarely do they attempt to define; far more commonly all we are offered is equivalents. The student will be aware of the pitfalls here for the user.

DICTIONARIES OF NAMES

The problems of nomenclature have already been touched on (page 41), but as guides through the jungle there are several dictionaries devoted solely to names: the coverage of H L Gerth van Wijk *Dictionary of plant names* (The Hague, Nijhoff, 1911-6) in two volumes, and of H N Andrews *Index of generic names of fossil plants, 1820-1950* (Washington, US Geological Survey, 1955) is world-wide, while that of R D MacLeod *Key to the names of British butterflies and moths* (Pitman, 1959) is obviously limited.

Some of these dictionaries endeavour to trace who it was who first coined each name, citing literature references where possible. Two classic and continuing works (both in English despite their Latin titles) are S A Neave *Nomenclator zoologicus* (Zoological Society of London, 1939-) which in the six volumes and quarter of a million entries so far published endeavours to provide 'as complete a record as possible of the bibliographical origins of the name of every genus and sub-genus in zoology'; and *Index kewensis* (Oxford, Clarendon Press, 1893-) with some half-a-million entries in over a dozen volumes, subtitled 'an enumeration of the genera and species of flowering plants . . . together with their authors' names, the works in which they were first published, their native countries and their synonyms'.

Of course many plants and animals are better known by their common name than their official name. Some dictionaries (*eg*, Gerth van Wijk above) do include such names, but there are also special dictionaries: C E Jackson *British names of birds* (Witherby, 1968), and R M Carleton *Index to common names of herbaceous plants* (Boston, Hall, 1959) are examples from each side of the Atlantic.

Names frequently encountered in everyday speech are those eponymous words like cardigan, sandwich, boycott, mackintosh that have entered the languages as the names of people with whom the things or practices they stand for were associated. Science and technology seem particularly prone to this habit, and the literature is dotted with terms like Ohm's law, Heaviside layer, Parkinson's disease. Medicine has so many of these (some as obviously venerable as Adam's apple

47

and Achilles tendon) that a policy of voluntary limitation has been agreed, and in some fields such as anatomy they are now forbidden words. Once again it is true to say that the ordinary subject dictionaries do include many of these terms, but there are also dictionaries specialising in eponyms: two in contrasting subjects are D W G Ballantyne and L E Q Walker *A dictionary of named effects and laws in chemistry, physics and mathematics* (Chapman and Hall, second edition 1961) and C P Auger *Engineering eponyms* (Library Association, 1965). They present a contrast in methods too: the entries in the former merely define, for example, what the Doppler effect is, but give no information about the man, no date, and no literature references; the relevant entry in Auger, on the other hand, not only says what the Wankel engine is, for instance, but cites two references, including Wankel's own paper, with a note of an available translation.

Raising similar problems are the many thousands of trade names encountered in science and technology. The complications here are such that trade names are accorded separate treatment in chapter 14.

DICTIONARIES OF ABBREVIATIONS

Any regular reader of the literature of science and technology will have found it plentifully sprinkled with abbreviations, like VHF (very high frequency), H_2O (water), $t\frac{1}{2}$ (radioactive half-life). Many can be looked up in the regular dictionaries, general and subject, and the student will already know of dictionaries specialising in abbreviations, but so extensive now is the practice that there are several dozen such dictionaries solely for science and technology, such as F A Buttress *World list of abbreviations of scientific, technological and commercial organizations* (Hill, third edition 1966). This is, however, an area of much confusion: Charlotte Schaler[1], for instance, has documented twelve methods of forming abbreviations.

For purposes of study it is useful to distinguish five categories of what may loosely be called ' abbreviations ':

a) *The contraction* is simply a shortened form of the original, such as in (inch), Chem Soc (Chemical Society), Al (aluminium), where part of the word stands for the whole. Sometimes it is taken from an alternative form of the original: lb, Ag are abbreviations of the Latin words for pound and silver. This is a common method of abbreviating the title of a scientific periodical, such as *Brit chem eng*, or *Arch*

1 ' Technical abbreviations and contractions in English ' *Journal of chemical education* 32 1955 114-7.

biochem biophys, and Ake Davidsson *Periodica technica abbreviata* (Stockholm, Petterson, 1946) was an early attempt to provide a dictionary of such abbreviations.

b) *The initialism* is an abbreviation comprising only the first letter(s) of the original term(s), as UV (ultra-violet), bp (boiling point), emf (electromotive force). Contractions and initialisms of this kind make up the bulk of entries in works like E B Steen *Dictionary of abbreviations in medicine* (Cassell, second edition 1963) and L W Wallis *Printing trade abbreviations* (Avis, 1960).

c) *The acronym,* a comparative newcomer, is a word consisting of the initials of a group of words, that is to say, a pronounceable initialism, for instance, laser (light amplification by stimulated emission of radiation), OSTI (Office for Scientific and Technical Information), ELDO (European launcher development organisation), scuba (self-contained underwater breathing apparatus). So popular and useful have acronyms become that it is now common practice so to contrive the name of a new device or project or material that its initials will form a neat acronym, *eg,* ADMIRAL (automatic and dynamic monitor with immediate relocation, allocation, and loading). Listing more than ten thousand, mainly from the fields of aviation and astronautics, is R C Moser *Space-age acronyms* (New York, Plenum Press, second edition 1969). The so-called ' telescope ' words like Mintech (Ministry of Technology), and Fortran (formula translation) are not acronyms according to strict definition, since they comprise more than simply the initials of the component words, but they are a particularly common form of abbreviation in German and Russian. Perhaps the most notorious is Gestapo (Geheime Staats Polizei, secret state police).

d) *The code* is the most difficult to define because it may or may not be an abbreviation and it may or may not be a pronounceable word. The point to grasp is that a code, whether a letter-code or word-code, may be assigned by a method other than abbreviation, perhaps even arbitrarily, such as TSR2 for a new military plane, or ADO 16 for a prototype car. Pluto, for instance, was the code-name during world war II for the system of cross-channel fuel-lines supplying the Allied invasion forces in Europe in 1944, but it was also the abbreviation (and indeed the initialism and acronym) for pipe line under the ocean. Apollo, on the other hand, the code-name for the US man-on-the-moon project, is not an abbreviation and not an initialism and not an acronym. Code-names are quite likely to result from arbitrary allocation: in military and security contexts the code-name may be deliberately selected as a 'cover' word, with the object of being as non-committal as possible, such as Barbarossa, for Hitler's invasion of Russia, but many others are quite fanciful choices, as the code-names given to typhoons by meteorologists, such as Edna, Lucy, Helen. A guide to some 8,500 of these is F J Ruffner and R C Thomas *Code names dictionary* (Detroit, Gale, 1963). An example of a guide to code-letters (as opposed to code-names) is H F Redman and L E Godfrey *Dictionary of report series codes* (New York, Special Libraries Association, 1962), which endeavours to identify the agency originating a particular US government research report from the code letters and numbers assigned.

e) *The symbol,* on the other hand, is never formed by abbreviation, and quite frequently is not in letter form at all. The plus and minus signs are probably the best known, and indeed mathematics abounds with symbols. Many of them, it is true, are in letter form, such as J (mechanical equivalent of heat), y (altitude), although in some cases the letter has been modified as in ℞ (recipe) and £ (pound sterling) and there are separate lists, *eg*, British Standards Institution *Letter symbols for electronic valves* (BS 1409:1950). Greek letter forms are commonly used: everyone is familiar with the symbol π for the ratio of the circumference of a circle to its diameter. But the most difficult of all to keep track of bibliographically are the non-literal, non-numerical symbols: some of them do have a tenuous connection with what they stand for (the biological symbol ♀ for female is said to be a stylised representation of the hand-mirror of Venus, Goddess of Love), but many are chosen without apparent relevance. There is fortunately a wide selection of guides, usually arranged systematically because of the impossibility of alphabetical order. The most comprehensive is O T Zimmerman and Irvin Lavine *Scientific and technical abbreviations, signs and symbols* (Dover, NH, Industrial Research Service, second edition 1949), but other examples are D D Polon *Dictionary of architectural signs and symbols* (New York, Odyssey, 1966), and R G Middleton *Electrical and electronic signs and symbols* (Indianapolis, Bobbs-Merrill, 1968). Attempts at standardisation have met with a substantial measure of success here: the British Standards Institution *Graphical symbols for general engineering* (BS 1553:1949-) and *Letter symbols, signs and abbreviations* (BS 1991:1961-) are but two of several similar titles, both still in progress. A systematic compendium of symbols approved as standards in the United States is Alvin Arnell *Standard graphical symbols: a comprehensive guide for use in industry, engineering and science* (McGraw-Hill, 1963). The compiler explains the significance of this work: ' To an engineer, the compact symbols of graphics represent the language of communication. This book is a dictionary of that language.'

There are also interlingual dictionaries of abbreviations, such as G E M Wohlauer and H D Gholston *German chemical abbreviations* (New York, Special Libraries Association, 1965) which gives the complete German form with an English equivalent. They are particularly needed for Russian, for ' No other language has given birth to abbreviations of its current scientific and technological terminology to the extent which modern Russian has and no other technical language is more saturated with acronyms ', according to Henryk Zalucki *Dictionary of Russian technical and scientific abbreviations with their full meaning in Russian, English and German* (Elsevier, 1968). The Library of Congress Aerospace Technology Division *Glossary of Russian abbreviations and acronyms* (1967) lists no fewer than 23,000 with English translations.

One last word of warning to the student: probably because it is so difficult for anyone (other than perhaps the compiler himself) satisfactorily to assess within a short time the worth of a dictionary, this category of work is probably more variable in quality than any other likely to be regularly consulted.

The user should note the words of W R Turnbull in the preface to his bibliography mentioned earlier: ' Like the thousands of other books being published each year, many are outstanding and the degree of back-breaking labour is evident on every page, while others are mediocre or almost totally inadequate '.

FURTHER READING

A L Gardner ' Technical translating dictionaries ' *Journal of documentation* 6 1950 25-31.

5

HANDBOOKS

The reference work most frequently consulted by the working scientist and technologist is the handbook. These compilations of miscellaneous information in handy form are of course found in other fields also, but they are seen at their best in science and technology, and the student should lose no time in examining for himself some of the classic titles. The *Standard handbook for mechanical engineers* (McGraw-Hill, seventh edition 1967) is still known as ' Marks ' after L S Marks, editor of the first five editions from 1916 to 1951. Similarly, ' Perry ' always refers to the *Chemical engineers' handbook* (McGraw-Hill, fourth edition 1963), edited from 1934 to 1950 by J H Perry. The amount of data contained in works of this kind is staggering: they both for instance, have about two thousand illustrations. And N A Lange and G M Forker *Handbook of chemistry* (McGraw-Hill, tenth edition 1967), *Machinery's handbook* (New York, Industrial Press, eighteenth edition 1968), and *Standard handbook for electrical engineers* (McGraw-Hill, tenth edition 1968) each has over two thousand pages.

To help the librarian understand the aim of these ' one-volume reference libraries ', as they have been called, it is only necessary to read what they have to say about themselves in their forewords and prefaces and introductions: O R Frisch *The nuclear handbook* (Newnes, 1958) is ' a day to day reference book'; E U Condon and H Odishaw *Handbook of physics* (New York, McGraw-Hill, second edition 1967) contains ' what every physicist should know '; D E Gray *American Institute of Physics handbook* (New York, McGraw-Hill, second edition 1963) is ' a working tool '. J H Potter *Handbook of the engineering sciences* (Van Nostrand, 1967) is more explicit about its role ' to assemble, categorize, and digest the more or less enduring fundamental considerations of the principal engineering sciences on a level approximating that of the first year graduate student in engineering '. And H H Huskey and G A Korn *Computer handbook*

(New York, McGraw-Hill, 1962) maintain that 'sufficient detail is presented so that anyone competent in the field can proceed to construct a computer'.

Handbooks are the first port of call when a straightforward factual problem arises in a particular subject field. It has been claimed that a library with no more than a sound collection of handbooks can answer 90 percent of quick-reference queries. As the last title indicates, the kind of information they contain is of a very practical nature: the *Steel designers manual* (Crosby Lockwood, third edition 1966) aims 'to bridge the gap between the normal textbook on the theory of design and its practical application in structural engineering . . . [and] provides authoritative data facilitating economic and efficient practice'. And in the words used in Cyril Long *Biochemists' handbook* (Spon, 1961), 'so far as possible, opinions and suggestions are not included'. One type of user frequently in the compiler's mind is the worker without access to a large literature collection. Indeed handbooks are often described as 'bench-books'.

Within one volume such works cannot hope to be comprehensive, but in any case that is not what the users want. What they seek is convenience combined with reliability: handbook editors try to provide this by the most skilful selection of data, the most time-saving arrangement and indexing, supported by regular (often annual) and painstaking revision. To become known as 'the chemists' Bible', the *Handbook of chemistry and physics* (Cleveland, Chemical Rubber Co, forty-ninth edition 1968) has had to maintain the highest standards since 1913. As might be expected, designed as they are for day-to-day consultation on the job, they do not usually give literature references, though there are notable exceptions: there are over six thousand in P L Altman and D S Dittmer *Biology data book* (Washington, Federation of American Societies for Experimental Biology, 1964), still known as 'Spector', after the editor of the first edition, W S Spector *Handbook of biological data* (Philadelphia, Saunders, 1956).

Although the titles so far mentioned in this chapter make up no more than a fraction of the total, it will have been noticed that they are preponderantly American, and that many of them come from the same publisher. This indeed is a precise reflection of the total picture, which is dominated by major American houses like McGraw-Hill, Wiley, and Van Nostrand. In Britain too, there are firms like Newnes and Iliffe with a sound reputation for their series of handbooks, *eg,* M G Say *Electrical engineer's reference book* (Newnes, twelfth edition 1968), E Drury *Builders' and decorators' reference book* (Newnes, [fifth edition, 1962]), L E C Hughes *Electronic engineer's reference book* (Iliffe, third edition 1967), F Langford-Smith *Radio designer's handbook* (Iliffe, fourth edition 1963). What has perhaps not been brought out so far is the extensive subject range of these handbooks. It is now common to find one or more for each major discipline or industry: representative titles are E F Kaelble *Handbook of x-rays* (McGraw-Hill, 1967), Illuminating Engineering Society *IES lighting handbook* (New York, fourth edition 1966), W I Orr *Radio handbook* (Foulsham, 1967), F S Merritt *Building construction handbook* (McGraw-Hill, second edition 1965), E E Grazda *Handbook of applied mathematics* (Van Nostrand, 1966), *Clay's public health inspector's handbook* (Lewis, twelfth edition 1968). The library user indeed is often surprised to find substantial handbooks devoted to quite specialised topics, *eg*, Universities Federation for Animal Welfare *The UFAW handbook on the care and management of laboratory animals* (Livingstone, third edition 1967), R T Liddicoat *Handbook of gem identification* (Los Angeles, Gemmological Institute of America, fifth edition 1957), V H Laughner and A D Hargan *Handbook of fastening and joining of metal parts* (McGraw-Hill, 1956).

The reader will have noticed that these works are not always called handbooks (or manuals, which is simply the Latin-derived term for the same thing): reference book is often used as a quite satisfactory alternative. What it is important to recognise is the handbook under some less revealing title: Mott Souders *The engineer's companion* (Wiley, 1966), Erich Rabald *Corrosion guide* (Elsevier, second edition 1968) and S Glasstone *Sourcebook on atomic energy* (Van Nostrand, third edition 1967) are all handbooks. On occasion the disguise is even more difficult to penetrate: despite its title, G Wyszecki and W S Styles *Color science* (New York, Wiley, 1967) is quite clearly a handbook by any definition; the preface describes it as a ' collection of concepts and methods, quantitative data and formulas ' and goes on to explain that ' descriptive and qualitative material . . . which would probably find a place in a textbook or treatise on color is not included '. The student should be warned, however, about the German word ' Handbuch ', frequently used, as in Joseph Durm and Herman Ende *Handbuch der Architektur* or S Flügge *Handbuch der Physik*. A dictionary would translate the term as handbook, thus providing a neat illustration of the limitations of dictionaries of equivalents, already cautioned against on page 47. These works are treatises, not handbooks: the former came out in 62 volumes between 1892 and 1907; the latter is still incomplete but so far has reached almost 70 volumes since 1955. And as the student will now have learned to expect, there are works described in plain English as handbooks or manuals that are really something else: the American Welding Society *Welding handbook* (Macmillan, sixth edition 1968-) in five volumes is more of a treatise than anything else, and L A Borradaile *Manual of elementary zoology* (OUP, fourteenth edition 1963) is a long-established textbook.

In fact, almost by definition a handbook (or manual) has to be in one volume if it is to stay ' handy '. Yet so vast and rapidly increasing is the store even of selected data that a worker ' ought to know ' that a number of well-known titles have had to extend to two, three, or even more volumes. *Specification: the standard reference book for architects, surveyors and municipal engineers* (Architectural Press) has appeared annually since 1898, but in recent years has become two volumes; though the third edition was a single volume, J G Cook *Handbook of textile fibres* (Watford, Merrow, fourth edition 1968) is now two; C J Smithells *Metals reference book* (Butterworths, fourth edition 1967) has grown from one to three volumes over eighteen years. And there are even larger works: S McLain and J H Martens *Reactor handbook* (New York, Interscience, second edition 1960-4) is in four volumes (in five) and the *Handbook of toxicology* (Philadelphia, Saunders, 1956-9) is in five volumes.

The layout adopted by handbook editors is almost invariably systematic, and a feature is the widespread use of the tabular form for the presentation of appropriate data. For satisfaction in use, of course, this arrangement demands adequate indexing, and in any assessment of the worth of a handbook this is an important aspect to consider. This is not to say, however, that other arrangements are never found: alphabetical order is used in G S Brady *Materials handbook* (McGraw-Hill, ninth edition 1963), in N I Sax *Dangerous properties of industrial materials* (Reinhold, third edition 1968) which used to be titled *Handbook of dangerous materials*, and in M Grieve *A modern herbal* (Cape, 1931).

This last example reminds us that there are a number of special categories of handbook-type reference works that are found in one discipline only. Because it is common practice in the building trade to obtain estimates before work commences, we find price-books giving up-to-date information on wage rates, prices of materials, methods of calculation, and a host of related matters. *Spon's architects' and builders' price book* is now approaching its hundredth edition, and *Laxton's building price book* (Kelly's Directories) passed the hundred mark before world war II.

A flora is a botanical handbook giving detailed information on the taxonomy (*ie,* classification) and distribution of plants in a defined area. There are popular, amateur floras such as the best-selling W Keble Martin *The concise British flora in colour* (Ebury Press, 1965), and scholarly, professional floras like A R Clapham *Flora of the British Isles* (CUP, second edition 1962) with *Illustrations* (1957-63) in four separate volumes. Floras may be local: W G Travis *Flora of South Lancashire* (Liverpool Botanical Society, 1963); or national: H W Rickett *Wild flowers of the United States* (McGraw-Hill, 1966-), to be in five volumes; or continental T G Tutin *Flora Europea* (CUP, 1964-) in four volumes. There are fortunately two helpful guides to the mass of titles: N D Simpson *A bibliographical index of the British flora* (Bournemouth, the author, 1960) with more than 65,000 entries, and S F Blake *Geographic guide to floras of the world* (Washington, US Department of Agriculture, 1942-) as yet incomplete. Similar compendia (which can be found in zoology also) are taxonomic indexes, species catalogues, check lists, and the like. A number of them (including some floras) are specially designed to aid the identification of biological material, often a far from simple task. Many textbooks also contain these keys, as they are called, *eg,* for fungi in G Smith *Industrial mycology* (Arnold, fifth edition 1960), and ' keying out a species ' in a flora or fauna key is one of the first skills the young biologist learns. A useful guide is J Smart and G Taylor *Bibliography of key works for the identification of the British fauna and flora* (British Museum (Natural History), Systematics Association, second edition 1953).

Pharmacopoeias are drug handbooks, particularly those with some degree of official sanction. The *British pharmacopoeia* (BP for short) appears every five years by the authority of the General Medical Council. The *British pharmaceutical codex* (BPC) appears simultaneously with BP and is published for the Pharmaceutical Society of Great Britain. Of less official standing (but perhaps of wider use) is *The extra pharmacopoeia (Martindale)* (twenty fifth edition 1967). Supplementing pharmacopoeias are formularies such as the *British national formulary,* which are virtually recipe books. Since drugs and their prescription are so closely controlled by law, it is usual for each country to have its own pharmacopoeia and related literature, although the *International pharmacopoeia* (Geneva, World Health Organisation, 1951-5) in two volumes with *Supplement* (1959) is an attempt at establishing international standards.

Formularies other than pharmaceutical make a category of reference book that could well be studied with handbooks. Distinctly practical in nature these are compilations of formulas with instructions for making particular products or producing certain reactions. The largest collection (80,000 tested formulas) is H Bennett *The chemical formulary* (New York, Chemical Publishing Co, 1933-), with thirteen volumes so far: sample recipes are for lipstick, rust-remover, kinky hair straightener, and powdered Worcestershire sauce. A handier one-volume work is *Henley's twentieth century book of formulas* (New York, new edition 1945). These are mainly for the amateur: for the professional there are works like R E Silverton and M J Anderson *Handbook of medical laboratory formulae* (Butterworths, 1961) or P Gray *The microtomist's formulary and guide* (Constable, 1954).

The data about the physical world that is the stock-in-trade of the handbook is constantly being refined, updated, or even superseded. Frequent revision is essential for any handbook that wishes to remain reliable. A limited number have chosen the loose-leaf method: B A Maynard *Manual of computer systems* (Gee, 1964-) in four volumes, Royal Institute of British Architects *Handbook of architectural practice and management* (1963-) in three volumes. Most pin their faith in regular new editions. There are several, indeed, that are undergoing continuous modification and are issued annually: *Radio amateur's handbook* (West Hartford, Conn, American Radio Relay League) has been appearing since 1926, and *Kempe's engineer's yearbook* (Morgan) since 1894. Although described as a yearbook, the latter is without doubt a handbook through and through, and indeed one of the most widely used and respected. In a number of other annual works, however, we have signs of the handbook merging with the yearbook, and even the directory, as in the ' *Metal industry* ' *handbook and directory* (Iliffe) and the *British instruments directory and data handbook* (United Trade Press). Directories and yearbooks are examined in chapter 6.

TABLES

As mentioned earlier, data in tabular form is a feature of many handbooks, and as the proportion of tables to text increases, the handbook as a form of literature merges into the book of tables. Because science, particularly the physical sciences, and technology are largely concerned with quantification, numerical information takes up much of the literature. Tables are a convenient way to present clearly details such as melting points, atomic weights, and solubilities. Indeed in certain fields like spectroscopy, or thermodynamics, or crystallography, tables are vital to the whole study and progress of the discipline, simply because such a large amount of information has been collected in tabular form, *eg,* R G J Miller and H A Willis *Irscot: improved structural correlation tables* (Heyden, 1966), J H Keenan and J Kaye *Gas tables* (New York, Wiley, 1948), International Union of Crystallography *International tables for x-ray crystallography* (Birmingham, Kynoch Press, 1952-62) in three volumes.

The purpose of tables is to save time. The information could of course be presented in a number of ways, or even left to lie where it was first reported in the primary sources. Indeed, as R T Bottle points out ' . . . many details of physical properties are deeply buried in the literature, and effort, patience and time are required to retrieve them. For this reason the standard books of tables are invaluable.' Provided, he might have added, that layout and indexes are designed with this prime aim in mind.

First to be considered are the great general exhaustive compilations such as the *International critical tables* (New York, McGraw-Hill, 1926-33) in seven volumes and an index. Two characteristics are plain: this is a collection of *critical* data, *ie,* the values given are those regarded as most reliable in the opinion of the three hundred expert consultants; secondly, full bibliographical references to the primary literature are given to enable the researcher to form his own estimate of the status and accuracy of the data. It is interesting to note that the language barrier is virtually non-existent in tabular compilations, especially if the explanatory text is multilingual, as here. In fact the great French and German compilations corresponding to *ICT* can be quite profitably consulted by a research worker with a minimum knowledge of the language.

For everyday use, however, what the worker needs by him is a selective listing, matched to his needs, and as up-to-date as possible. One of the best examples, first published in 1911, is G W C Kaye and T H Laby *Tables of physical and chemical constants* (Longmans, thirteenth edition 1966). A larger American compilation with over nine hundred tables, although still in a handy single volume is the

Smithsonian physical tables (Washington, Smithsonian Institution, ninth edition 1954). An even more selective listing is W H J Childs *Physical constants selected for students* (Methuen, eighth edition 1958). Though there will probably always be a demand for handy, time-saving books of tables such as these, the practicability of the comprehensive compilations like *ICT* is now in question. Not only are they difficult to keep up-to-date and cumbersome to use, but the editorial problems of digesting the vast amount of new data that accumulates daily are virtually insuperable by conventional means. The trend in recent years has been towards a more fragmented approach to the compilation of tables, with separate works devoted to particular sets of values or designed for particular disciplines, *eg*, A Seidell *Solubilities: inorganic and metal organic compounds* (Van Nostrand and American Chemical Society, fourth edition 1958-66), W Kunz and J Schintlmeister *Nuclear tables* (Pergamon, 1963-), S Eilon *Industrial engineering tables* (Van Nostrand, 1962). As a direct result of this piecemeal approach, however, two new difficulties have arisen: first, how to avoid wasteful duplication of effort; and second, how to keep track of the many tables being published.

As the mass of data has grown, books of tables have appeared covering more and more limited areas, but even so, some of these works represent an investment of many thousands of man-hours in calculating and manipulating and setting out. In many cases the conventional methods of type-setting have proved too slow or too expensive or both, and the tables have been printed from typescript by photolitho-offset, *eg, Organic electronic spectral data* (Interscience, 1960-), in several volumes which covers the literature from 1946. For some, mechanisation has been seen as offering a partial solution to the problem of absorbing the new and improved values reported in the primary literature. Now available on subscription are several services which furnish data on a continuing basis in the form of punched cards, *eg, Documentation of molecular spectroscopy* (Butterworths) with some two thousand edge-punched cards each year.

The computer in particular has proved its worth in this struggle. It has been used for a number of years now to manipulate data: the National Engineering Laboratory *Steam tables 1964* (Edinburgh, HMSO, 1964) were partly calculated by computer, and D P Jordan and M D Mintz *Air tables* (New York, McGraw-Hill, 1965) were ' photographically reproduced from the original printed computer output '. More recently it has been grasped that for many purposes (*eg*, reference as opposed to current awareness) data need not be printed out and published at all, provided that it is on call when required. Computer technology now permits this facility through remote-access terminals, with the computer memory acting as a ' data bank '.

The tables so far mentioned have been mainly concerned with physical and chemical data, but these are not the best known. Mathematical tables of logarithmic values, trigonometrical functions, square roots and so on are far more widely used: indeed many of us are familiar with four-figure tables from our schooldays. Even simple tables such as these are ten times as accurate as the common ten-inch slide rule used for engineering calculations, but scientists and technologists often need more extended tables: a good standard set is the two-volume *Chambers's six-figure mathematical tables* (1948-9), although it is possible to go further, *eg*, with A J Thompson *Logarithmetica Britannica: being a standard table of logarithms to twenty decimal places* (CUP, 1952). But tables of this kind are only one part of the scene: there are hundreds of others. Some publishers have extensive series that are just tables, *eg*, ' Pergamon mathematical table series ' in some forty volumes. A useful collection is R S Burington *Handbook of mathematical tables and formulas* (McGraw-Hill, fourth edition 1965), and there are a number of bibliographies, of which the outstanding example is Alan Fletcher *An index of mathematical tables* (Oxford, Blackwell, second edition 1962). An interesting feature of this two-volume work is a 150-page list of known errors in specific published tables.

Mechanisation has proved useful for mathematical tables also: the series of *Royal Society mathematical tables* from volume 11 (1964) are in the form of a ' photographic reproduction of sheets prepared on a card-operated typewriter '. One effect of the computer in this field has been the great increase in the number of tables: ventures such as F W Kellaway *Penguin-Honeywell book of tables* (Harmondsworth, 1969) would have been impossible without the computer, for not only have all the standard tables been recalculated, but the actual pages of the book have been typeset by computer, the first Penguin to be so produced.

There are many other fields of endeavour where tables are widely used, and many other kinds of data that benefit from tabular presentation. Three examples will suffice: *World weather records, 1951-60* (Washington, US Weather Bureau, 1965) is the latest in a continuing series going back to the earliest observations; J H Kenneth and G R Ritchie *Gestation periods: a table and a bibliography* (Commonwealth Agricultural Bureaux, third edition 1955); the *Admiralty tide tables* appear annually in three volumes.

Finally the student should be aware of one difficulty that is found whenever any value is expressed in numerical units, such as inches, or ounces, or degrees Fahrenheit, or gallons. It is a fact that not everyone uses the same units. Universal metrication is seen as the long-term solution, but in the meantime workers have to rely on aids such as conversion tables. There are official tables published by bodies such as the British Standards Institution *Conversion factors and tables* (BS 350:1959-62), as well as long-established works like O T Zimmerman and I Lavine *Conversion factors and tables* (Dover, NH, Industrial Research Service, third edition 1961). An interesting multilingual work with explanations in five languages is Stephen Naft *International conversion tables* (Cassell, revised edition 1965) which claims to have many tables found nowhere else, and is also available in a condensed edition for schools, universities, and colleges, as *Cassell's concise conversion tables* (revised edition 1965).

6

DIRECTORIES AND YEARBOOKS

Directories are basically lists of names and addresses, arranged for reference purposes in a variety of ways to match the requirements of their users and frequently updated: the annual *'Packaging'* directory is a list of manufacturers arranged alphabetically, for instance, but the *'Contract journal'* directory arranges its list of building and other contracts by county. The *Architect and contractors yearbook*, on the other hand is arranged in classified order. To this basis of name and address is often added a range of other information about, for example, the products of a company (as in the *BEAMA directory* from the British Electrical and Allied Manufacturers' Association), the activities and staff of an international organisation (as in the *World nuclear directory*), or the academic qualifications of a scientist (as in the *Directory of British scientists*).

The word ' directory' will not necessarily appear in the title: two very well-known examples are *Ports of the world* and *Scientific and learned societies of Great Britain*. Very commonly, where a directory appears annually, it will be called a yearbook, even though it may not at all fit the strict definition of yearbook (see page 70): the *Mining year book* (annually since 1887), *Oil and petroleum year book* (annually since 1910), and *Year book of technical education* (annually since 1957) are all quite clearly directories.

It is staggering to contemplate the range of subjects dealt with by directories: even the most specialised industries seem to be covered, as for example, in the *Waste trade manual and directory*, the *Turkeys year book*, and the *Directory of shoe machinery*. In fact, directories make up the largest single category of reference books: a large library might have as many as four thousand current directories. Fortunately there are bibliographical guides available: in addition to the general lists by Henderson and Klein with which the student will be familiar there are guides specifically for science and technology: the most comprehensive is A P Harvey *Directory of scientific directories* (Hodgson, 1969). A much slighter 'provisional checklist' is the Library of Congress Science and Technology Division *Directories in science and technology* (Washington, 1964).

TRADE OR INDUSTRIAL DIRECTORIES

As the directory is primarily an instrument of commerce with its main aim the bringing together of buyer and seller, it is not surprising that the majority of directories fall into this category. Because industry and trade is international, many such directories attempt world-wide coverage: the *International plastics directory* covers 76 countries, *Skinner's wool trade directory of the world* claims to be the only one of its kind in the field, and the *Tanker directory of the world* is an interesting example of a guide to a rapidly-growing international industry. But the majority of titles are national in scope: well-established and well-used examples for the student to examine are the *Timber trades directory* (first published 1892), *British plastics year book* (first published 1931), *Chemical industry directory* (first published 1923), and *Ryland's coal, iron, steel, tinplate, metal, engineering, foundry, hardware and allied trades directory* (first published 1881 and understandably known simply as ' Rylands ').

A number of these directories have sharpened their focus by calling themselves ' buyers' guides ', although of course all trade directories are by definition guides for buyers (or sellers). They are not only guides to commodities and manufactured goods but increasingly to services also. Not all of them are recent innovations: *Where to buy everything chemical* has been going for half a century, ' *Machinery's* ' *annual buyers' guide* for almost as long.

A feature of this category of directory that the student may encounter is the habit that has grown up from casual and obvious beginnings of calling a particular work the ' red book ' (as with the *Rubber red book*) or the ' green book ' (as with *British tractors and farm machinery . . . the green book*) or the ' blue book ' (as with the *Electrical and electronics trades directory: the blue book*). This can be a source of confusion in a library when an engineer asking for the ' blue book ' may not realise that the *Rubber directory of Great Britain*, the *Advertiser's annual*, and of course, the *Hotel and catering blue book*, the *Brewery age blue book*, the *Leather and shoes blue book*, the *Printing trades blue book*, *Davidson's textile blue book*, and *MacRae's blue book . . . the buying directory for engineering* are all

63

known to their respective trades or professions simply as the ' blue book '.

The observant student will also have noticed that many of these industrial directories are published by or have close ties with a particular periodical, as ' *The aeroplane* ' *directory of British aviation*, or the ' *Chemical week* ' *buyers' guide*. These may be produced and priced quite separately from the periodical itself, as the ' *Electrical and electronic trader* ' *yearbook*, or more commonly may be issued as part of the subscription, as the ' *Chemistry and industry* ' *buyers' guide* each autumn, or the *Computer directory and buyers guide* which actually comprises the June number of *Computers and automation*.

What could perhaps be regarded as a special category within this general group of industrial or trade directories is the exhibition catalogue. An example to study is the annual *Handbook of scientific instruments and apparatus*, which is a guide to the year's Physical Society exhibition and virtually a directory and buyers' guide.

Trade names will be discussed in chapter 14, but it is worth noting that many industrial directories contain lists of trade names, and that there are examples of directories devoted entirely to them, *eg*, ' *The ironmonger* ' *directory of branded hardware*, a list of over 40,000 names, and the *Food trade directory of trade marks and trade names* with about 12,000.

DIRECTORIES OF INDIVIDUAL SCIENTISTS AND TECHNOLOGISTS
It is commonly the case that directory-type information (name, address, etc) is called for about a person rather than a firm. As we have seen, information of this kind is sometimes found in the trade directories, *eg*, the ' *Control* ' *industry guide and digest* contains a ' who's who ' of control engineers, and *Instruments, electronics, automation, year book and buyers guide* lists computer personnel. There are also a large number of separately published professional directories: one of the largest is the *International directory of research and development scientists* (Philadelphia, Institute for Scientific Information, 1968) with over 125,000 names and addresses. It is more usual to find examples confined to one particular discipline, although still on a world-wide or international basis, *eg*, the *World directory of crystallographers* (International Union of Crystallography, fourth edition in preparation) which lists over four thousand specialists from over fifty countries, the *World directory of mathematicians* (Bombay, Tata Institute, second edition 1961), covering 71 countries, or O D Sherby

Directory of research metallurgists and metal physicists in Western Europe (Office of Naval Research, 1958); more frequently they list the experts within a particular country or area, *eg*, the *Directory of official architects and planners*, or the *Sci-en-tech register* of 15,000 Chicago scientific and technological personnel.

The most frequently encountered of such lists, however, are the membership directories of the scientific and technological societies and institutions, *eg*, the annual Institution of Mechanical Engineers *List of members* in three volumes, the biennial Royal Institute of Chemistry *Register of corporate members*, and the *Combined membership list* of some 25,000 members of the American Mathematical Association, the Mathematical Association of America and the Society for Industrial and Applied Mathematics. These are particularly common in the United States and sometimes attain a vast size, *eg*, the American Chemical Society *Directory of membership* with over 90,000 names. Frequently such lists appear as part of a yearbook, *eg*, in the Institute of Agricultural Engineers *Year book*, or in the Royal Institution of Great Britain *Record*. Less commonly they are found in a periodical, *eg*, the Society of Plastics Engineers annual listing in the *Journal*. In certain professions such lists are statutory, *eg*, the *Medical register*, issued annually under the authority of the General Medical Council and the *Register of patent agents* maintained by the Chartered Institute of Patent Agents on behalf of the Board of Trade.

The value of such tools in facilitating communication between scientists and the technologists can well be imagined, but there is one type of professional directory which has a further function, *ie*, the list of consultants. This is to some extent also a trade directory insofar as it serves to draw to the attention of possible clients the services that the consultant is selling, *eg*, the Royal Institute of Chemistry *Directory of independent consultants in chemistry and related subjects* (fourth edition 1961).

These works are not really biographical sources in the full sense of that term, since the information given about each individual is usually biographically meagre, often no more than name, academic or professional qualifications, post held, and address. They are primarily location tools, and unlike the biographical dictionaries, which are almost invariably alphabetically ordered, these professional directories sometimes prefer a geographical arrangement, as in the *List of research workers . . . in agriculture, animal health, and forestry* published every three years by the Commonwealth Agricultural Bureaux, or even more commonly, a geographical index to the main alphabetical sequence, as in the third volume of the Institution of Mechanical Engineers *List* mentioned earlier, or in the Royal Institute of British Architects *Directory*. Another arrangement encountered occasionally is by subject, with the specialists listed under their speciality, but once again this approach is more widespread in the form of the subject index to the main alphabetical sequence, as in the *Register of incorporated photographers*.

Once the entries begin to include more than mere directory information, as in the *Medical list,* this type of reference work begins to merge into the biographical sources discussed in chapter 20.

DIRECTORIES OF SCIENTIFIC AND TECHNOLOGICAL ORGANISATIONS
Including all kinds of organisations other than industrial firms (which are covered by trade directories) these make up a large and important class. Their special significance for the information worker lies in the fact that they are (with the professional directories) the guides to the non-documentary sources of information described in chapter 1 (pages 15-6), such as government departments, learned societies, universities, etc. These sources by definition are not part of the literature of science and technology, but the directories and guides to them are, and therefore receive attention here.

It is a truism to say that science knows no frontiers, and there are indeed a large number of organisations of scientists and technologists that are international in scope and which should perhaps be considered first. At the summit of world science sits the International Council of Scientific Unions, comprising bodies such as the International Union of Pure and Applied Chemistry, the International Union of Biological Sciences, and many others. A useful guide to study is their *Year book* published in Rome by the ICSU Secretariat. Much wider in scope though now out of date is the UNESCO *Directory of international scientific organizations* (Paris, second edition 1953) covering 264 bodies in science and technology. More recent is the OECD *International scientific organizations: catalogue* ([Paris 1965]) and its *Supplement* (1966). A particularly valuable tool for librarians since it concentrates on some 449 of these organisations particularly as sources of information is the Library of Congress General Reference and Bibliography Division *International scientific organizations: a guide to their library, documentation, and information services* (Washington, 1962).

A common variant of this type of directory is the research guide: as most research in science and technology nowadays is institutional rather than individual, lists such as C H Williams *European research index* (Hodgson, second edition 1969) are virtually directories of organisations of all kinds where research is conducted, or promoted, *eg*, government research establishments, independent research institutes, university research departments, and research laboratories of industrial firms.

Of course most scientific organisations are national in scope, and are described in directories such as the National Academy of Sciences *Scientific and technical societies of the United States* (Washington, [eighth edition] 1968), or *Industrial research in Britain* (Harrap, sixth edition 1968), or the Battelle Memorial Institute *Directory of selected scientific institutions in the USSR* (Columbus, Merrill, 1963). An ambitious recent attempt at world coverage is the *Guide to world science* (Hodgson, 1958-) planned to be in twenty volumes. Aiming to ' unravel this organisational web of science ', each volume takes the form of a directory of important scientific establishments of all kinds within a single country (or group of countries). The last volume is devoted to international scientific organisations, both intergovernmental and non-governmental.

All the directories mentioned so far under this heading have covered the whole of science and technology, but there are many examples concerned with particular subject fields. A guide of international scope is the *Aerospace research index* (Hodgson, fourth edition 1969); a national listing is the Royal Society *Marine science in the United Kingdom: a directory of scientists, establishments and facilities* (1968).

There are also directories, international and national, of particular types of scientific and technological organisations. The Ministry of Technology *Technical services for industry* (1967) is in effect a directory of government research stations, grant-aided industrial research associations, and government departments concerned with science and technology. By way of contrast, *Industrial research laboratories of the United States* (Washington, Bowker, twelfth edition 1965) is devoted to non-governmental institutions. The *Research centers directory* (Detroit, Gale, third edition 1968) concentrates on yet another type of organisation: university-sponsored and other non-profit research bodies in the United States and Canada. Two unique directories are Patricia Millard *Trade associations and professional bodies of the United Kingdom* (Pergamon, fourth edition 1968) and UNESCO *World directory of national science policy-making bodies* (Paris, 1967-) to be in four volumes.

Of particular significance to the librarian are directories of scientific and technological information organisations, such as libraries, specialised information centres, data banks, etc. Again lists are found of international and national scope. The UNESCO *World guide to science information and documentation services* ([Paris, 1965]) covers 65 countries and will be complemented by a similar volume for technology; the OECD *Guide to European sources of technical information* (Paris, second edition 1964) describes five hundred institutes in 19 countries. National lists concentrating mainly on conventional libraries are the *ASLIB directory: volume one, information sources in technology and commerce* (1968) and A T Kruzas *The directory of special libraries and information centers* (Detroit, Gale, second edition 1967), but the *Directory of federally supported information analysis centers* (Washington, Committee on Scientific and Technical Information, 1968) deals solely with specialised information centres, and the Library of Congress National Reference Center for Science and Technology *A directory of information resources in the United States: physical sciences, biological sciences, engineering* ([Washington, 1965]) is an example of a list designed to cover all possible sources. Its alphabetical arrangement makes an interesting contrast with the classified order of the Battelle Memorial Institute *Specialized science information services in the United States: a directory* (Washington, National Science Foundation, 1961). New types of information organisations stemming from the introduction of mechanisation are listed in directories such as OSTI *Data activities in Britain* (third edition 1969) and the *Directory of computerized information in science and technology* (New York, Science Associates, 1968). So many indeed are these directories that we have for some years had a bibliography in the field: *Directories of science information sources: international bibliography* (The Hague, FID, second edition 1967) lists 360 titles from 58 countries.

Because science and technology have as their sphere of action the physical world, a feature of research is the use made of collections of physical objects such as rocks, insects, anatomical specimens, trees, micro-organisms, etc. In certain fields such as biology these assemblages have especial value for purposes of identification. The guides to the existence of these collections form a small but interesting category of directories, *eg,* UNESCO *Directory of meteorite collections and meteorite research* (Paris, 1968) which covers 49 countries, R A Howard and others *International directory of botanical gardens*

(Utrecht, International Bureau for Plant Taxonomy and Nomenclature, 1963), and the *International zoo yearbook*, which lists zoos and aquaria of the world, are all international in coverage. D H Kent *British herbaria* (Botanical Society of the British Isles, 1957) and M M Roberts *Public gardens and arboretums of the United States* (New York, Holt, 1962) are obviously national. Museums are listed in two periodical articles: 'Museums of science and technology', *Museum* 20 1967 150-228; and an earlier article with an identical title, *Technology and culture* 4 1963 133-47.

Of course, as well as serving as non-documentary sources of information, many of the organisations listed in these directories have a major role as producers of the literature, particularly of the primary literature such as research periodicals and conference papers, and also of bibliographical sources such as abstracting services and reviews of progress. This aspect of their work will be given due attention in the appropriate later chapters.

In proportion to its numbers the directory is probably found less frequently in its pure form than any other reference book, *ie*, as a list of names and addresses. Very commonly indeed are directories extended by the addition of, for instance, handbook-type information or encyclopedia- or dictionary-type information. The '*Gas journal*' *directory*, for example, commences with a section of over seventy pages entitled 'Handbook of technical data'; the *Finishing handbook and directory* contains a lengthy alphabetically arranged 'Encyclopaedia of finishing'; the *Carpet annual* includes a dictionary of trade terms in English, French, German and Italian. Biographical and statistical information is frequently found also, as in the *Chemical industry directory and who's who*, and the *Cotton year book* respectively. A feature is sometimes made of special articles on particular aspects of the subject, as for instance in the *Fisheries year-book and directory*. Particularly valuable for librarians are the bibliographies often included, as the list of 'Trade journals of interest to the fishing industry' in the last example, or the 'Best farm books' in *The Farmer's yearbook*, or the 'Comprehensive list of specialized weekly and monthly journals' in *The fruit annual*.

Carrying as they do all this 'subject' information, such directories are clearly not merely tertiary sources (see page 15). It is obvious that we have here again the merging of categories noticed so frequently before. This is even more evident in such omnibus compilations as the *Trader handbook*, which is subtitled 'a legal, technical and buying guide for the motor, motor cycle, and cycle trades'. It is probably best to regard 'directories' as a convenient term covering a heterogeneous range of very useful reference books, each to be taken on its merits, but needing particularly close examination to determine its uses. And while not forgetting the strict definition of the term, the student should not pay too much heed to the way a work is described in its title: both the *Brewery manual* and the *World radio and TV handbook* are pure directories!

YEARBOOKS

It is useful to know how to recognise a yearbook proper, for many 'yearbooks' are nothing more than directories published annually, as we have seen. The student will know from his previous studies that a yearbook's concern is the events of a particular year. An obvious example is the *McGraw-Hill yearbook of science and technology*: its prime function of course is to supplement the *McGraw-Hill encyclopedia,* but it is in the form of an account of the year's developments in the field. Another common type of yearbook is the statistical: both the *Sugar year book* and the *Minerals yearbook* are mainly given over to statistics of production, consumption, imports, exports, stocks, etc. And then there are a limited number of works concerned with the forthcoming year (or years), as for example the *Yearbook of astronomy*, which predicts the phases of the moon and the planetary orbits and the like.

In the various disciplines of science and technology much of the task of digesting the developments of the year is performed by annual reviews of progress which will be considered at length in chapter 11. It should be noted, however, that there are examples of works which do try to combine the function of annual review and yearbook and directory, *eg, Computer yearbook and directory.*

And finally, there is at least one very famous 'yearbook' that is neither a yearbook nor a directory: the *Yearbook of agriculture* of the United States Department of Agriculture. In fact it is not even a reference book, strictly speaking. Each annual issue deals at length with a particular topic such as 'Climate and man' (1941), 'Insects' (1952), 'Seeds' (1961), 'Outdoors USA' (1967).

7

BOOKS 'IN THE FIELD'

All the previous chapters in this volume have been about reference books. Now of course any work, from the Bible and Shakespeare to the daily newspaper, can be used for reference, can be 'looked up' for a specific point, but the touchstone of a true reference book is that it is deliberately arranged to facilitate such consultation. The titles discussed in this chapter are not: they comprise the works on the subject, the books 'in the field'. They serve to expound or to systemise or to discuss or to reveal their subject. The forms they take most often are the treatise, the monograph, and the textbook.

TREATISES

Like the encyclopedia, the treatise attempts to cover the whole of its subject field, but in most other ways it differs significantly. The encyclopedia aims to furnish a concise survey of each topic it treats, intelligible (at least initially) to the non-specialist; it tries to provide first and essential facts only and does not attempt to exhaust the subject;[1] and it is arranged (usually alphabetically) so that its contents are easy of access. The true treatise, on the other hand, sets out to be exhaustive, aiming for a complete presentation of the subject with full documentation. The only information likely to be excluded is elementary or introductory material. The classic work by J W Mellor *A comprehensive treatise on inorganic and theoretical chemistry* (Longmans, 1922-37) in sixteen volumes with *Supplement* (1956-) describes itself as 'A complete description of all compounds known in inorganic chemistry'. L M Hyman *The invertebrates* (McGraw-Hill, 1940-) claims 'The intent of this treatise, then, is to furnish a

[1] As has been pointed out earlier (page 30), the large comprehensive encyclopedias of the Kirk-Othmer type are not typical, sharing as they do something of the character of the treatise.

reasonably complete and modern account '. J A Steers *The coastline of England and Wales* (CUP, second edition 1964) describes itself as ' a systematic approach to the whole subject . . . wider in treatment than the local monographs . . . more comprehensive than the existing works . . . This book is a physiographical treatise '. Works of this kind are not for the beginner, although concessions are sometimes made as in G M Ramachandran *Treatise on collagen* (Academic Press, 1967-) where ' The editors have attempted to make the treatment extensive, for those who are comparatively new to the subject, as well as intensive, so that the latest developments are discussed in particular detail for those specializing in each field '. But not all such works are so academic: describing itself as ' a comprehensive treatise ' is the two-volume George Ellis *Modern practical stairbuilding and handrailing* (Batsford, 1932), ' primarily a practical book for workshop use ', and still not superseded.

The arrangement of a treatise is invariably systematic: its function is to set out its subject in all its aspects, not to serve as a source of quick reference. The five-volumed J R Partington *Advanced treatise on physical chemistry* (Longmans, 1949-54) has volumes dealing in turn with the properties of gases, of liquids, and of solids. The work by Mellor mentioned above follows the order of the periodic table. A revealing light is cast on the contrasting roles of the treatise and encyclopedia by I M Kolthoff and P J Elving *Treatise on analytical chemistry* (New York, Interscience, 1959-), planned in three parts, each in a dozen or more volumes. Part I concerns itself with theory and practice, part II with the analytical chemistry of the elements, but when the editors came to treat the analytical chemistry of industrial materials in part III they found it was not possible to present this area with the same systematic and critical treatment attempted in parts I and II. They decided that an encyclopedic treatment would do better justice to the requirements of the practising analytical chemist: this treatment is now appearing as *Encyclopedia of industrial chemical analysis* (Interscience, 1966-).

As has been described (page 13) science progresses by the accumulation of facts derived from observation and experiment, and these researches are reported first in the primary literature. The main function of the treatise is to impose some sort of system on the information scattered through time and space in these primary sources. In the preface to G T Rado and Harry Suhl *Magnetism: a treatise on modern theory and materials* (New York, Academic Press, 1963-) we read that 'The need for a consolidation of almost all theoretical and experimental aspects of magnetically ordered materials is the motivation for the present work'. And the editors of Marcel Florkin and E H Stotz *Comprehensive biochemistry* (Elsevier, 1962-), to be in thirty volumes, write: 'Beyond the ordinary textbook the subject matter of the rapidly expanding knowledge of biochemistry is spread among innumerable journals, monographs, and series of reviews. The Editors believe that there is a real place for an advanced treatise in biochemistry which assembles the principal areas of the subject in a single set of books.'

A corollary of this function of digesting the primary literature is, of course, the full bibliographical references that are a feature of treatises. Herein lies their strength as sources of information: for the exhaustive approach to a problem (see page 18) they provide an excellent point of departure. Even more valuable is the treatise which is critical, that is to say which discusses the status and worth of the material presented: Mellon regards this as the highest type of treatise.

The subject scope of a treatise is broad rather than narrow: indeed it is by virtue of its wide range that it is able to perform its valuable task of synthesis and consolidation. But what constitutes 'broad' is a matter of opinion: there are treatises on what some might regard as narrow topics, *eg*, the three-volume G C Ainsworth and A S Sussman *The fungi: an advanced treatise* (Academic Press, 1965-), or the two-volume 'comprehensive treatise' by R Houwink and G Salomon *Adhesion and adhesives* (Elsevier, second edition 1965-7). On the other hand, some subjects are too extensive for such treatment: it is generally agreed that it is impossible to compile a general treatise in the field of applied chemistry that would remain reliable for long, and the practice has been to produce works devoted to a particular area within the field, *eg*, plastics.

Indeed, it is this vast and insoluble problem of the time-lag that could spell the end for the treatise. By its very nature, it is first of all a slow and painstaking task to compile a treatise, and the beginning and the end quite often span a generation or more: the 78-year-old author of the treatise on *The invertebrates* noted above announced her retirement after volume 6, and the work is to be continued by others. Secondly, by virtue of its systematic form it is very difficult to keep up-to-date. And yet scientists have always felt the need for a work that would include everything about their subject. There is, however, substantial evidence (in physics, for example) that the relative importance of the treatise is decreasing: its place is being taken by the monograph and the textbook.

MONOGRAPHS

The monograph resembles the treatise in many ways, and there are a number of examples where the two categories overlap. In the General Introduction to perhaps the best known current series, the American Chemical Society monograph series, appears the explanation: '. . . in the beginning of the series, it seemed expedient to construe rather broadly the definition of a Monograph. Needs of workers had to be recognised. Consequently among the first hundred Monographs appeared works in the form of treatises covering in some instances rather broad areas.' As an example could be instanced the four-volumed Fred O'Flaherty *The chemistry and technology of leathers* (Chapman and Hall, 1956-65).

Traditionally the main difference between the treatise and the monograph is that in contrast to the broad subject scope of the treatise the field of the monograph (as its Greek origins indicate) is a narrowly-defined single topic. This distinction is demonstrated neatly in a treatise like H F Mark *Man-made fibers: science and technology* (Interscience, 1967-8) which is in fact a *collection* of monographs, each by different authors on different topics, such as 'Acrylic fibers ', ' Nylon 66 ', ' Fiber testing '. Indeed one way to regard a monograph is as one section of a hypothetical treatise, although this is not the complete picture. Typical monographs are A J C Andersen and P N Williams *Margarine* (Pergamon, second edition 1965), E L Wheeler *Scientific glassblowing* (New York, Interscience, 1958), R P Wodehouse *Pollen grains: their structure and significance in science and medicine* (Hafner, 1959), and R W Barton *Telex: a detailed exposition of the telex system of the British Post Office* (Pitman, 1968).

Within its limited subject field, however, the monograph strives to be comprehensive. Mellon quotes the example of Dorsey's work on *Properties of ordinary water substance* which ' devotes 673 pages to perhaps our best known chemical '. Wilfred Francis *Coal: its formation and composition* (Arnold, second edition 1961) and G L Walls *The vertebrate eye* (Hafner, 1963) both have over eight hundred pages. And an even more striking example is J Z Young *The anatomy of the nervous system of octopus vulgaris* (OUP, 1969), thirty years in preparation, with over four hundred pages and seven hundred illustrations, which 'describes the nerve cells and tracts of the the brain of the octopus as completely as possible with present knowledge '.

Like the treatise, the monograph attempts the same systematizing approach to its subject: the preface to Z E Jolles *Bromine and its compounds* (Benn, 1966), a volume of just under a thousand pages, states: 'One of the aims of this monograph is to bring together in accessible form and logical order, under one cover, the factual knowledge of the chemistry and technology of bromine, for the benefit of the advanced student and the research worker '. The preface to Peter Alexander and R F Hudson *Wool: its chemistry and physics* (Chapman and Hall, second edition 1963) speaks of a great need for a ' comprehensive monograph . . . in view of the great diversity of the sources of information which include the journals of chemistry, physics and biology, and the applied journals of the textile industry '.

Full documentation is usual with the monograph, as with the treatise: P S Lawson *Tobacco: experimental and critical studies: a comprehensive account of the world literature* (Baltimore, Williams and Wilkins, 1961), is ' compiled from more than 6,000 articles published in some 1,200 journals . . . the resultant monograph is primarily directed to the research worker and specialist ', and the bibliography takes up 110 of the 944 pages.

It would be a mistake to assume, however, that it is only the narrowness of the topic it treats that distinguishes the monograph from the treatise. A prime feature of the monograph is its emphasis on contemporary knowledge: although comprehensive in aim, it will not normally include the detailed historical or background material found in a treatise. P G Forrest *Fatigue of metals* (Pergamon, 1962), a work of over four hundred pages with a bibliography of 686 references, illustrates this well in its preface: '. . . no comprehensive British book on fatigue, as distinct from reports of conferences, has been published during the last 30 years. It therefore seemed that there was a need to provide a general account of the present knowledge.' Perhaps it was this kind of monograph that the great Dr Johnson had in mind when he declared ' there should come out such a book every thirty years, dressed in the mode of the times '. It is this emphasis on contemporaneity that partly accounts in some fields of endeavour for the increased reliance on the monograph (or rather series of monographs) at the expense of the treatise. It is obviously far easier for an editor or publisher to ensure that individual monographs are kept up-to-date than to revise a massive treatise. In rapidly developing subjects the attempt to produce even a monograph can prove abortive: in the preface to H A Liebhafsky and E J Cairns *Fuel cells and fuel batteries* (Wiley, 1968) we read that ' since we began this book five years ago, the world has spent a hundred million dollars, more or less, on research and development aimed at fuel cells and fuel batteries . . . Consequently we changed course: what was envisioned in 1962 as a complete account became a historical and expository guide.'

The monograph plays a further role that is not shared by the treatise: often it can be seen to stand midway between the ' repository' function of the treatise and the ' reporting' function of the periodical. It is not unknown for a monograph to contain previously unpublished material, and instances can be found where it serves to bypass the conventional primary sources of publication such as periodicals and research reports. More commonly, however, the monograph serves in a handy form to draw to the attention of working scientists and technologists the research results reported, perhaps rather obscurely or inaccessibly, in the primary literature. The preface to M Szcepanski *The brittleness of steel* (Wiley, 1963) well illustrates this: ' In 1952, when work on this book was begun, there was not a single book dealing with steel brittleness, although the scientific and economic significance of this problem was well recognised. On the other hand, the number of papers on this subject was so great that a person seeking specific information was lost.' The student should not conclude from all this that a monograph is a mere compilation or digest. To begin with, the skill required to assess just what is of contemporary concern within a particular field demands an author who is an authority in the subject: the critical approach is essential in a good monograph. Furthermore, the particular role that the monograph can play in stimulating ideas, crossing disciplinary barriers, and pointing the way forward, depends upon an author who can write interesting and readable prose.

It will not have escaped attention that many monographs appear in series under a general editor or editorial board but with individual authors, and most of the major scientific and technological publishers have one or more. Examples of titles in such series are W N Christiansen and J A Hogbom *Radiotelescopes* (1969) in the Cambridge monographs on physics; G V T Mattheus *Bird navigation* (second edition 1968) in the Cambridge monographs on experimental biology; E A Lynton *Superconductivity* (third edition 1969) in Methuen's monographs on physical subjects. As with other forms of literature, the publishers are sometimes learned societies and institutions, *eg*, the Carnegie Institution of Washington monographs, or the Carus mathematical monographs of the Mathematical Association of America.

As a description of a category of information source the term monograph is perhaps less abused than some others, but the student will need to be on his guard against claims such as that on the jacket of R J Bray and R E Loughhead *Sunspots* (Chapman and Hall, 1964): ' the first comprehensive treatise on sunspots to be published '. The authors' preface is more modest, pointing out that the work deals with ' just one aspect of solar activity, namely sunspots ', and describes itself more accurately as a ' monograph '. A more frequent occurrence is the use of the term monograph for a work which really falls short of full monograph status: the excellent little guide by Frank Newby *How to find out about patents* (Pergamon, 1967) refers to itself as a monograph nine times in the two-page preface, but lacks the comprehensiveness of coverage and fullness of documentation expected of a true monograph.

The student should also be aware of the use of the term monograph in its original sense, where it was confined to biology. In this field it still is used for an account of a single species, genus, or larger group of plants, animals or minerals. Such monographs may be no more than a few pages in length, and many of them are never separately published: D H Kent *Index to botanical monographs* (Academic Press, 1967) lists nearly two thousand of these from nearly four hundred periodicals.

TEXTBOOKS

A textbook is a teaching instrument: its primary aim is not to impart information about its subject but to develop understanding of it. It concentrates on demonstrating principles rather than recounting details. These principles are supplemented by descriptions, explanations, examples, etc, only in sufficient number to ensure the reader's grasp of the principles. Since the essence of teaching is selection, a textbook may not cover the whole field of its subject, and may not even describe the latest practice, unless there is some new principle involved. The material included is representative rather than comprehensive. Although each of the following three textbooks are substantial volumes, they are deliberately selective: A D Imms *A general textbook of entomology* (Methuen, ninth edition 1957) admits ' we have not attempted systematically to cover the literature after 1952 ';

F A Cotton and Geoffrey Wilkinson *Advanced inorganic chemistry: a comprehensive text* (Interscience, second edition 1966) emphasises that ' it is intended to be a *teaching* text and not a reference book ', despite its 1,148 pages; W H Salmon and E N Simon *Foundry practice* (Pitman, revised edition 1957) speaks of other books on the subject as ' highly technical and cluttered up with a mass of theoretical material . . . not infrequently they approach the subject less as practical guides than as academic expositions '. This selective approach is evident also in the the bibliographies included in textbooks: these are most frequently in the form of suggestions for further reading rather than exhaustive lists of references. And of course many textbooks are without bibliographies—unthinkable in a monograph or treatise.

A feature of textbooks is their longevity. Simply because of their concentration on principles, good textbooks will continue to sell to generation after generation of students, even in the more rapidly changing disciplines within science and technology. Of course the better examples are kept up-to-date by frequent new editions, but apart from matters of detail there is often not a great deal of difference between one edition and the next. This is understandable: the success of a textbook depends not on its worth as a source of up-to-the-minute data but on whether its method of presentation enables its users to learn about the subject. Examples still going strong that have demonstrated their worth in this way are H Cotton *Applied electricity* (Macmillan, sixth edition 1966) first published 1951, A W Judge *High speed diesel engines . . . an elementary textbook* (Chapman and Hall, sixth edition 1967) first published 1933, K Newton and W Steeds *The motor vehicle: a textbook* (Iliffe, eighth edition 1966) first published 1929, *Samson Wright's applied physiology* (OUP, eleventh edition 1965) first published 1926, *Fream's elements of agriculture* (Murray, fourteenth edition 1962) first published 1892, and most famous of all, *Gray's anatomy* (Longmans, thirty fourth edition 1967) first published 1858.

On the other hand, textbooks are subject to the particular hazard of changing methods of teaching, that is to say not so much a change in content as in the selection of the content and the way it is presented. Good textbooks take account of this: H Cotton *Principles of electrical technology* (Pitman, 1967) is indeed a new book, but it is virtually the seventh edition of a similar textbook by the same author adapted in accordance with the changing approach to electrical technology, and containing more on electronics and less on machines than its predecessor. Inorganic chemistry is another subject that has seen major changes over the last few years in the way it is taught as can be seen by comparing the emphasis upon atomic structure and chemical bonds in F A Cotton and G Wilkinson *Advanced inorganic chemistry* (Interscience, second edition 1966) with the more conventional approach of N V Sidgwick *The chemical elements* (OUP, 1950). An interesting example of a book changing its character as its subject develops is Oscar Faber and J R Kell *Heating and air-conditioning of buildings* (Architectural Press, fourth edition, 1966): its preface relates that ' When this book first appeared in 1955 as a series of articles in *The architect's journal*, literature on the subject was almost non-existent . . . Now the mass of information and literature is bewildering . . . A fresh bibliography is published every month. Thus although originally this book supplied a need otherwise unfulfilled, it must now be regarded more as in the nature of an introduction to the subject: each chapter could well form the basis of a separate book, the whole becoming an encyclopaedia. But in such form it would no doubt be beyond what the average reader requires.'

A more easily perceived change in an increasing number of textbooks, particularly from United States publishers, is the emphasis on colourful and attractive physical presentation. S N Namovitz and D B Stone *Earth science: the world we live in* (Van Nostrand, second edition 1960) is a striking example of colour printing, with an illustration on almost every page. High quality illustrations are a feature of Karl von Frisch *Biology* (Harper and Row, 1964), a translation of a German school text. The most obvious indication of this bright new approach are the full colour illustrations on the front and back boards of works of this kind. Titles worthy of study include W H Hayt and G W Hughes *Introduction to electrical engineering* (McGraw-Hill, 1968), H A Guthrie *Introductory nutrition* (St Louis, Mosby, 1967), G C Beakley and H W Leach *Engineering: an introduction to a creative profession* (Collier-Macmillan, 1967), M D Potter and B P Corbman *Textiles: fiber to fabric* (McGraw-Hill, fourth edition 1967), and J H Dubois and F W John *Plastics* (Reinhold, 1967).

In some instances, this increased attention paid to book design is matched by a revolutionary approach to teaching, as for instance in the texts that have emerged from the Biological Sciences Curriculum Study in the United States and the Nuffield Science Teaching Project in Britain, *eg, Biological science: molecules to man* (Arnold, 1963), and *Nuffield chemistry* (Longmans, 1966-), issued in several parts.

To the observant the textbook and the monograph may appear similar, for both are simply books 'in the field', and neither is immediately distinguishable by alphabetical order (as is a dictionary), or by tabular arrangement (as are many handbooks) or by type of data included (as is a directory). Yet even a superficial examination of the reading habits of scientists and technologists shows that monographs and textbooks are used quite differently. It is for this simple reason that it is vital for the information worker who wishes to advise on their use that he grasp clearly the basic differences. One of these has been discussed: the selective nature of the textbook. Two others are equally significant. In the first place, it is only rarely that the author of a textbook is presenting the results of his own researches: his book is usually put together from the contributions others have made to the subject. He may not even be an authority on the topic: indeed most textbooks cover so wide an area that no man could be a real authority over the whole field. The author's role is that of teacher, scanning the whole area to select his examples, and arranging them to illustrate the underlying principles he has chosen to expound. In the words of Mellon, ' . . . the author's chief function is to select, arrange, and discuss '.

The other significant difference to note about textbooks is that they are written at several levels, quite precisely calculated, although this is not always obvious without examination of the author's preface. P J Durrant *General and inorganic chemistry* (Longmans, third edition 1964), for instance, is 'for the use of students in their last year at school or their first year in a technical college or university'; A I Vogel *A textbook of quantitative inorganic analysis* (Longmans, third edition 1961) is 'to meet the requirements of University and College students of all grades'; and Malcolm Dixon and E C Webb *Enzymes* (Longmans, second edition 1964) is a work ' dealing with the general principles of the subject at the research level'. Some texts are even more particularly tailored for specific courses: A J Grove and G E Newell *Animal biology* (UTP, seventh edition 1966) is 'for the requirements of candidates for the General Certificate of Education at Advanced Level'; T D Eastop and A McConkey *Applied thermodynamics for engineering technologists* (Longmans, 1963) is ' written to meet the needs of students studying Applied Thermodynamics as part of a Higher National Certificate and Higher National Diploma courses in Mechanical Engineering '; and D C T Bennett *The complete air navigator* (Pitman, seventh edition 1967) is subtitled 'covering the syllabus for the Flight Navigator's Licence'. Such titles can sometimes be made obsolete by the frequent changes that courses undergo.

The librarian should not be too ready to dismiss the textbook as of little account compared with the monograph. Not only has it a very different role in the pattern of scientific communication, but it is often more difficult to write than a monograph, which is usually produced by an expert for his fellows. It is sometimes the case that the expert is the last man to write a textbook, if all he possesses is a knowledge of the subject. Since the textbook is a work of instruction, ideally the author should be familiar with modern teaching methods, should have an understanding of the learning process, and be acquainted with the needs of students. It is little wonder that the typical textbook author is a teacher. Indeed, many textbooks have their origin in a course of lectures: the author of L M Milne-Thomson *Theoretical aerodynamics* (Macmillan, fourth edition 1966) says that the work is 'based on my lectures . . . at the Royal Naval College'; Therald Moeller *Inorganic chemistry: an advanced textbook* (Wiley, 1952) is 'the outgrowth of a one-semester lecture course' at the University of Illinois; and, forming an instructive contrast, C S G Phillips and R J P Williams *Inorganic chemistry* (Oxford, Clarendon Press, 1965-6) in two volumes is 'based on a one-year course of lectures which we have given over a number of years to Oxford undergraduates'.

It is highly unlikely that teachers who thus set down the content of their courses are planning their own redundancy: such texts are obviously intended not as substitutes for attending a course but as aids to its successful pursuance. They include detail that time does not permit the lecturer to include in class, or illustrations that can best be presented in book form, and provide the student with a permanent record of the exposition of the subject. At the very least, they can save the students time, as is explained in the preface to Robert Smith and H W Heap *Sheet metal technology* (Cassell, 1964): 'As teachers of the subject the authors have found that the students have to spend valuable time taking notes. It was to avoid this and to enable the students to utilize their time more advantageously, that prompted the writing of these books'. And at least one textbook has ventured to reproduce in facsimile a teacher's own handwritten lecture notes: James Holmes *Manuscript notes on weaving* ([Burnley, 1917?]). There are examples of textbooks (more commonly in science and technology than in the humanities), however, that do try to provide the whole of a course within their covers: R Passmore and J S Robson *A companion to medical studies* (Oxford, Blackwell, 1968-), to be in three volumes, claims that ' the student will find in it more than sufficient information to enable him to pass his examinations '.

It so happens that medicine is one profession that it is not possible to enter by private study, but there are many disciplines where students need not attend a course, and for these the textbook has an even more vital service to give. It is probable that the majority of such students use the normal type of textbook we have been discussing, but in recent years there has been an explosion of interest in self-instruction. Not only are there many more examples of conventional text designed for this purpose such as J E Thompson *Mathematics for self-study* (Van Nostrand, third edition 1962), in five volumes, and K N Dodd *Teach yourself analogue computers* (EUP, 1969), but we now have large numbers of programmed texts, both linear and branching. An example of the former is RCA Service Company *Fundamentals of transistors: a programmed text* (Prentice-Hall, 1966), and of the latter G M Berlyne *A course in renal diseases* (Oxford, Blackwell, 1966).

It is possible to distinguish a number of other variants from the standard textbook. Very frequently encountered is the book of problems or questions with solutions or answers: two contrasting examples from the USSR and the USA are N M Belyayev *Problems in strength of materials* (Pergamon, 1966) and the long-established J L Hornung and A A McKenzie *Radio operating questions and answers* (McGraw-Hill, thirteenth edition 1964). Less common is the ' reader ', which collects in one volume a number of selected extracts from other books, periodicals, etc. The preface to C S Coon *A reader in general anthropology* (Cape, 1950) explains its function: '. . . there are too many students for our library. Most of the books we would like to have the students read are out of print, and we could not buy enough copies to go around even if we had the money and the library space . . . There is only one practical solution—a reader.' The work contains twenty selections, some over forty pages in length, from monographs, learned society transactions, reports of scientific expeditions, lectures, historical accounts, classic Greek authors, etc.

Perhaps the most common variation on the basic textbook is the ' how-to-do-it ' book, in its various manifestations. Clearly these are works of instruction, but they aim at more than simply inculcating principles or developing understanding. With their aid the student should find himself able to *do* something, usually of a practical kind, *eg, Whittaker's Dyeing with coal-tar dyestuffs* (Bailliere, sixth edition 1964), or George Sideris *Microelectronic packaging: interconnection and assembly of integrated circuits* (McGraw-Hill, 1968). Naturally such works include less theory and more practice than conventional textbooks, but like them (and unlike monographs) they are written for students at various specified levels: E B Bennion *Breadmaking: its principles and practice* (OUP, fourth edition 1967) is for students preparing for the examinations of the City and Guilds of London Institute; The Admiralty *Naval marine engineering practice* (HMSO, 1959-) is a training text for engineering mechanic ratings.

Compared to the theoretical textbooks, these practical works are more frequently found directed at the actual worker at the bench or in the field rather than the student in the classroom. R W Castle *Damp walls: the causes and methods of treatment* (Technical Press, 1964) and U Langefors and B Kihlstrom *The modern technique of rock blasting* (Wiley, second edition 1967) are quite obviously practical texts. In other cases, it is made explicit in the book itself: the opening words of the preface of A Davidsohn and B M Milwidsky *Synthetic detergents* (Hill, fourth edition 1967) are ' This is a work by practical men for practical men '; J D Long *Modern electric circuit design* (McGraw-Hill, 1968) is ' written for those individuals who would design circuits as an occupation '; and in their preface John McCombe and F R Haigh *Overhead line practice* (Macdonald, third edition 1966) write of their ' hope that it will be read by foremen and linesmen, as well as by engineers and others associated with the transmission and distribution of electricity by overhead lines '. A number are in the form of compilations of tried and tested methods, ' cook-books ' of science and technology in fact, *eg,* R G J Miller *Laboratory methods in infra-red spectroscopy* (Heyden, 1965), and P S Diamond and R F Denman *Laboratory techniques in chemistry and biochemistry* (Butterworths, 1966). Instruction books (often published by the manufacturer) for particular models of tractors, microscopes, computers, etc, make up a similar and extensive category of texts (see page 201).

A number of textbooks of this kind are indeed substantial and high-grade expositions of a particular skill: for instance, R H Davis *Deep diving and submarine operations* (St Catherine Press, seventh edition 1962) is an amazing compilation of underwater lore. But R G

Harry *The principles and practice of modern cosmetics,* volume I (Hill, fifth edition 1962), volume II (second edition 1963) and Charles Rob and Rodney Smith *Operative surgery* (Butterworths, second edition 1969-), to be in fourteen volumes, are instances where perhaps the description ' textbook ' is inadequate. The word ' practice ' commonly appears in titles of works in this category, *eg,* Victor Serry *British sawmilling practice* (Benn, 1963), or Ernest Davies *Traffic engineering practice* (Spon, second edition 1968), but again such works are often more than mere textbooks, as is made plain by the title of W O Skeat *Manual of British water engineering practice* (Cambridge, Heffer, third edition 1961). In fact these are examples of a familiar indeterminate type of work in the field of technology, part textbook (for study) part handbook (for reference), and even part monograph. Two other well established titles are J H Thornton *Textbook of footwear manufacture* (Heywood, third edition 1964), which the foreword claims as ' the first book to cover the whole range of footwear manufacture ', and Horace Thornton *Textbook of meat inspection* (Bailliere, fifth edition 1968), ' covering the whole field of meat hygiene in a simple yet comprehensive manner '.

Of course the field of ' how-to-do-it ' books is a happy hunting ground for the amateur enthusiast, and thousands of titles are produced specifically for the non-professional car mechanic, TV repairman, carpenter, and so on, *eg,* G W Mackenzie *Acoustics* (Focal Press, 1964) for ' the amateur hi-fi enthusiast ', and P H Smith *Tuning for speed and tuning for economy* (Foulis, third edition 1967). There are at least two bibliographies devoted to works in this category: *How-to-do-it books: a selected guide* (New York, Bowker, third edition 1963) lists over 4,000 relating primarily to spare-time activities, and F S Smith *Know-how books* (Thames & Hudson, 1956) likewise concentrates on practical texts for the general reader.

For the librarian textbooks raise a number of special problems of which the student should be aware. Most imperative is the frequently expressed view that textbooks have no place in a library. As works of instruction, designed for intensive study, they should be in students' hands as their personal possessions. The normal limited loan period of most libraries (imposed of course with the aim of ensuring fair shares) is inappropriate for a book which a student may need by him throughout a course. And as a source of information, the average textbook supplies nothing that is not available elsewhere, usually in more convenient form. A further difficulty for the librarian is the vast number of similar publications, particularly in those subjects which attract large numbers of students. Very often one of these works differs from the next only insofar as it represents another teacher's view as to how the topic should be expounded. Perhaps even more serious are the criticisms of unreliability in matters of fact, or misleading over-simplification, or lack of awareness of modern developments, that are levelled at textbooks, particularly school textbooks, from time to time in the scientific press. And yet, as the student will also be aware, textbooks are found in large numbers in many kinds of scientific and technological library.

INTRODUCTIONS AND OUTLINES

Although as categories these are less precisely defined, they are often classed with textbooks and are therefore briefly considered here. An introduction is clearly a *first* book in a subject, designed to lay the groundwork for its user, and leading on to a more advanced or detailed or particular study, *eg,* H R Broadbent *An introduction to railway braking* (Chapman and Hall, 1969). They are not necessarily for 'students': the two-volumed P W Abeles *An introduction to prestressed concrete* (Concrete Publications, 1964-6) is for 'civil engineers, architects, and contractors'. A number of them are planned, in the words of the foreword to B E Waye *Introduction to technical ceramics* (Maclaren, 1967), so as to be 'suitable for students and others', *eg,* M S Ghausi and J J Kelly *Introduction to distributed-parameter networks with application to integrated circuits* (Holt, Rinehart and Winston, 1968), which is 'for seniors and first-year graduate students as well as for self-teaching by practicing [sic] research engineers'. And of course a large number of 'introductions' are simply textbooks, *eg,* P Grosberg *An introduction to textile mechanisms* (Benn, 1968), 'aimed in general at the ordinary and honours degree student', C E Littlejohn and G F Meenaghan *An introduction to chemical engineering* (Chapman and Hall, 1959), 'intended primarily as a beginning text for students'; and P E Gray *Introduction to electronics* (Wiley, 1967), 'intended to be used in introductory or first-level courses'. Not all textbooks are introductions, on the other hand: a number of them are exceedingly advanced, *eg,* V I Smirnov *A course of higher mathematics* (Oxford, Pergamon, 1964), used in the USSR for 45 years, and filling six volumes in the English translation.

An outline covers the *whole* of its particular subject, but not in detail. Only the salient features are emphasised. Its aim is not so much to develop understanding (as the textbook), but to map out an area. Where a textbook (or an introduction) is designed for continuous study, and arranged on the assumption that it will be worked through in sequence, the outline can also be used quite easily for reference. The preface to E N Simons *An outline of metallurgy* (Muller, 1968) reflects this approach precisely when it speaks of the need for a book that 'without going into excessive detail, would give a general view of the subject, making it unnecessary to read an entire shelf-full of textbooks in order to grasp what is involved'. It is of more than minor importance for the librarian to be able to distinguish categories such as these. It can be seen that outlines (and to a less extent introductions) are particularly useful for the enquirer seeking to orientate himself in a field relatively new to him—the background approach, in fact, as described in chapter 1 (pages 17-8).

This chapter, and the previous chapters, have been in the main concerned with books. Now it is often said, somewhat scornfully, that in science and technology books contain only second-hand information. This, of course, is true. All the types of literature we have so far considered have been either secondary or tertiary sources: we have not yet studied the literature of science and technology at its heart, at its fountainhead. But the student should beware of underestimating the role of books. The chemist is probably as fully aware of the primary literature as any man but in the classic guide by E J Crane (see page 24 above) we read: 'Chemical books have most important uses. They introduce the novice to the general field of the science or of some part of it, explain new theories in the light of already known facts, and help to coordinate and systematize knowledge. They furnish information, exhaustive or not, in a form adapted for quick reference, and guide the searcher back to the original source by means of citations. Historical works record the development of the science, popular books initiate the public into its mysteries and elicit interest and support, and treatises on the various fields of chemistry give the reader the benefit of the long experience or combined researches of many workers. Who shall say that the chemist can depend on journals alone? The mere fact that nearly two thousand new books of chemical interest are published annually proves the demand for them.'

8

BIBLIOGRAPHIES

Before we leave for the time being our study of the literature of science and technology in book form, we should pause to examine those sources of information that are consulted to ascertain what books are indeed available on a subject—the bibliographies, in fact. Of course such lists of books are only one form of bibliography: as will become evident in succeeding chapters, all the various non-book materials such as periodicals, patents, theses, etc, have their own bibliographies (or indexes as they are often termed). Very common also is the mixed bibliography, listing books and non-book materials together irrespective of form.

Bibliographies of course are tertiary sources (see pages 15-6), and as such are particularly the librarian's province. Indeed, as has been mentioned, many of them are actually compiled by librarians, and it is from librarians that they obtain much (perhaps most) of their use. Traditionally, they are divided into current and retrospective, but for the student this can lead to confusion, for what is current today can be retrospective tomorrow: there are many examples of current bibliographical listings which do later serve their turn as retrospective bibliographical repositories, particularly when cumulated, *eg*, the monthly *ASLIB book list* forms the basis for *British scientific and technical books, 1935-52* (1956) and its supplements.

A more illuminating line for the student to pursue is to try to match the bibliographies he is studying to the various approaches made to them as sources of information on what books exist. Whether the users are scientists, technologists, or librarians and information officers, they demonstrate in their use of bibliographies precisely those ' approaches ' already observed in the general literature, ie, the Current, the Everyday (and Background), and the Exhaustive (see pages 16-8).

THE CURRENT APPROACH

If the available lists are any indication, keeping abreast of current books in science and technology is very much the librarian's preserve:

the two longest-established are both primarily annotated book selection tools, compiled by and for librarians and information officers, namely the *ASLIB book list* (1935-) and *New technical books* (1915-). The latter is a production of the New York Public Library, which reminds us that the accessions lists of appropriate libraries are widely used for the same purpose, allowing all to reap the advantage of the selection skills of others, *eg*, *The recorder*, a 'selected list of recently acquired publications' issued twice a month by Columbia University Science and Engineering Libraries, the University of London Library *Accessions list: Section VI—Science medicine, agriculture, engineering,* and the rather more specialised National Lending Library for Science and Technology monthly *List of books received from the USSR and translated books.* A particularly useful example worth singling out is *Lewis's quarterly list,* which is a little unusual in that it lists 'new books and new editions on sale and added to the medical, scientific and technical, lending library, which is of course a commercial library. Not all accession lists are separately published: the excellent annotated list of additions to the library of the Institution of Mechanical Engineers appears in the monthly *Chartered mechanical engineer.* Similarly a number of booksellers specialising in science and technology issue regular lists of recently published books in their field. A number of periodicals too make a feature of lists of new books: perhaps the most widely used is the extensive 'Recent scientific and technical books' pull-out section in the last issue for each month of *Nature.*

Many of these regular lists in periodicals are there simply to acknowledge copies of books sent for review by publishers, and surveys have shown that the book review columns in the scientific and technical press are one of the most important sources for the practising scientist of current information on new books. Of course, many periodicals carry such reviews, *eg*, *Engineer, Textile recorder, Bioscience, Chemistry in Britain, Science,* and thousands of others, but in some journals they extend to twenty, thirty, or even more pages in each issue, *eg*, *Physics today, Science progress, School science review.* Indeed, in recent issues of the *Quarterly review of biology,* for instance, the 'New biological books' department has reached over seventy pages, or over half the journal. And since 1965, in *Science books: a quarterly review,* published by the American Association for the Advancement of Science, we have a journal consisting entirely of reviews. Unique in this field is the monthly *Technical book review index* (also compiled by librarians), which under each book listed gives extracts from published reviews, and, as its title states, serves as an index to the book reviews which have appeared in over 2,500 journals. While not exactly furnishing book reviews in the accepted sense, many abstracting services do include announcements of new books, often with indicative abstracts of their contents, *eg, Chemical abstracts, Hosiery abstracts*.

A remarkable but unfortunately short-lived project in recent years was *Sci-tech book profiles* (and its companion *Medical book profiles*), which 'profiled' some 150 books each month by reproducing from each in reduced facsimile anything up to twenty of the more significant pages, *eg*, title page, table of contents, preface, list of contributors, index, etc.

It is also possible for the alert librarian to keep himself informed about books not yet published. Apart from ensuring he is on the mailing lists of the major scientific and technical publishers and that he receives the monthly batch of scientific titles from PICS (Publishers' Information Cards Services), he can refer to sources such as the invaluable feature 'Books to come: science, technology, medicine and business', which appears quarterly in the *Library journal*, where as many as a thousand forthcoming titles are listed with succinct and informative annotations.

THE EVERYDAY (AND BACKGROUND) APPROACH

Whether by the practitioner or by the librarian, this approach takes the form of a search for a suitable book on a particular subject of interest. The emphasis here is on appropriateness to the task in hand, as the need for information usually arises in the course of daily work. If the requirement is verbalised (as it has to be if the librarian is asked to help), it is usually for 'a good book on the subject', or 'the standard work', or even 'the best source'. The bibliographies of most immediate use here are the selective, annotated, evaluative lists (as opposed to the comprehensive, more purely 'bibliographical' lists used for the exhaustive approach, described below). A classic example is the work known universally as 'Hawkins' after its first editor: *Scientific, medical, and technical books published in the United States of America: a selected list of titles in print with annotations* (Washington, National Research Council, second edition 1958). This stands in the same sort of relationship to the current listing *New technical books* as does its British counterpart *British scientific and technical books, 1935-52: a select list of recommended books* (ASLIB,

1956) and its supplements to the *ASLIB book list*. Though technically ' selective ' rather than comprehensive, these are both substantial bibliographies with 8,000 and 11,000 titles respectively, and find their main use as tools for book selection and readers' advisory work in libraries (despite their need for updating). Serving a similar function is the very different *McGraw-Hill basic bibliography of science and technology* (1966), issued as a supplement to the *McGraw-Hill encyclopedia,* but subtitled ' recent titles on more than 7,000 subjects '.

A clearly discernible category within this class of bibliographies of basic books is the compilation specifically designed to suggest a representative list of titles for library purchase. Examples (in order of descending size) are Northeastern University (Boston) *A selective bibliography in science and engineering* (Boston Hall, 1964), ' a working collection for an undergraduate library ' of 15,000 titles; John Crerar Library (Chicago) *Industrial technical library: a bibliography* (Washington, Office of Industrial Resources, [1960]), ' annotated listings of approximately 3,000 books and periodicals, representing a sound guide for the selection of a balanced industrial technical library '; ASLIB *Select list of British scientific and technical books* (fifth edition 1957), about 1,300 items, without annotations, and limited to books in print; Carnegie Library of Pittsburgh *Science and technology: a purchase guide for branch and small public libraries* (1963), 1,034 annotated titles, updated by annual supplements; Association for Science Education *Science books for a school library* (Murray, fifth edition 1968), 650 titles. Aimed also at purchasers, though not necessarily libraries, is of course the regular British trade bibliography *Technical books in print* (Whitaker) with some 15,000 titles.

Yet another distinguishable type of selective list is the ' reader's guide '. Two established titles in this category are the National Book League *Science for all: an annotated reading list for the non-specialist* (revised edition, 1964), ' prepared in consultation with the British Association for the Advancement of Science ', and its precise transatlantic counterpart, H J Deason *A guide to science reading* (New York, New American Library, [1966]), produced by the American Association for the Advancement of Science.

THE EXHAUSTIVE APPROACH

As pointed out in chapter 1, ' exhaustive ' in this context is a relative term, for even where total coverage of the literature is the aim, a search is usually curtailed at the point where the law of diminishing returns begins to operate. Nevertheless, this is the area of activity where bibliographies really come into their own: to locate a bibliography of the subject is often the vital first step in the pursuance of a research project. Indeed, George Sarton, probably the best-known bibliographer of the history of science, and himself by training a mathematician, has stated that 'Each investigation must begin with a bibliography and end with a better bibliography'. Although the number and range of bibliographies already published frequently astound even the librarian, it is often necessary, however, for the investigator to compile his own if he finds himself the first in the field.

Such a worker, systematically seeking to trace all the books published in his subject would probably look first for the comprehensive retrospective bibliographies covering the whole field of science and technology. He would soon discover that such lists do not exist (for English titles at any rate) as separate publications. *American scientific books* (New York, Bowker) used to be 'a complete record of scientific and technical books published in the United States of America, and in four volumes covering 1960 to 1965 cumulated some 4,000 titles per year from the monthly *American book publishing record*. Since then, however, these scientific and technical titles are incorporated into the general annual listing *BPR cumulative* in a fashion similar to the *British national bibliography*. As the student will be aware, subject access is simplified to the relevant titles in both these general works (and their cumulations) by their arrangement in Dewey Decimal Classification order.

In point of fact, the most comprehensive broad-based retrospective bibliographies of science and technology are the published catalogues of the great scientific and technological libraries, *eg*, British Museum (Natural History) Library *Catalogue of the books* . . . (1903-40) in eight volumes; Wellcome Historical Medical Library *A catalogue of printed books* (1962-), of which the first volume lists 7,000 books printed before 1641; and Harvard University *Catalogue of the Library of the Museum of Comparative Zoology* (Boston, Hall, 1967) with 153,000 entries in eight volumes. Particularly worthy of study on account of its form is this last example: unlike the others, which are conventionally printed, it is made up of facsimiles of the actual cards from the catalogue in the library, reproduced by photolitho-offset, several to the page.

Printed library catalogues such as these have a particular value as bibliographies, for not only do they provide information about the existence of a book and date, publisher, format, etc, but they locate an actual copy by way of a bonus. A grave disadvantage with many works of this kind, however, is the lack of subject access: the three examples quoted, for instance, are all arranged alphabetically by author, with no subject index. All the other bibliographies mentioned in this chapter, both current and retrospective, are arranged by subject (or have a subject index): indeed the various 'approaches' so far described demand such access. But of course to produce a subject catalogue calls for much more intellectual effort than an author list, and until recently good examples in the science and technology field were few: the twin Science Museum Library short-title lists *Books on engineering: a subject catalogue* (HMSO, 1957) and *Books on the chemical and allied industries: a subject catalogue* (HMSO, 1961) contain together about 10,000 books arranged by the Universal Decimal Classification; a particularly handy list is the *Catalogue of Lewis's medical, scientific and technical lending library* (1965-6), of almost 40,000 titles, with an author/title sequence in one volume and the subject index in the other.

What has revolutionised the situation in the last few years has been the extensive use of the photolitho-offset method described above to reproduce in book form the card catalogues of several dozen of the world's major libraries, many of which are subject lists of one form or another. The largest in the field of science and technology, with well over a million books and pamphlets, is the John Crerar Library (Chicago) *Catalog* (Boston, Hall, 1967) in 77 volumes, 35 of which make up the *Author-title catalog,* and 42 the *Classified subject catalog* (with subject index). A contrasting arrangement is seen in the *Dictionary catalog of the National Agricultural Library, 1862-1965* (Boston, Hall, 1968-) to be in 68 volumes; in the *Classed subject catalog of the Engineering Societies Library* (Boston, Hall, 1963-), in 13 volumes and annual supplements, it is the author approach that is denied.

As has already been pointed out, this chapter concerns itself with bibliographies of *books;* yet in an exhaustive search, by definition, books make up only one part. And it is an observable fact that many published bibliographies are ' mixed ', inasmuch as they combine the features of a bibliography (*ie,* of books) and an index (*ie,* of periodical articles, research reports, patents, etc). They are themselves, as often as not, the published results of an exhaustive search of the literature. Each of the following, for instance, covers pamphlets, articles, and other printed material as well as books: G S Duncan *Bibliography of glass (from the earliest records to 1940)* (Dawsons, 1960) has over 15,000 references; H S Brown *A bibliography of meteorites* (Chicago UP, 1953) goes back as far as 1491 for its first entry; and A M C Thompson *A bibliography of nursing literature, 1859-1960* (Library Association, 1968) claims to be a contribution to history rather than a working tool. On occasion, however, special circumstances may dictate the exclusion of periodical articles for instance, even in a comprehensive bibliography: E A Baker and D J Foskett *Bibliography of food* (Butterworths, 1958) does so because ' they are easily traced by searching the abstract journals '; H C Bolton *Select bibliography of chemistry, 1492-[1902]* (Washington, Smithsonian Institution), with 18,000 titles included, would obviously be too vast an undertaking if not confined to separately published works.

It is easy to understand why this kind of specific subject bibliography is so highly regarded by the research worker obliged to carry out an exhaustive literature search during the course of a project. His particular problem is to find out whether such a search in his particular field has been carried out before; or, in practical terms, whether there is a bibliography. Here the librarian is particularly well placed to help, with his acquaintance with sources of information, and in particular with those intriguing tools, the bibliographies of bibliographies. He will of course be thoroughly familiar with the general tools, but he should know that there are a limited number of such titles in science and technology also, *eg*, C J West and D D Berolzheimer *Bibliography of bibliographies on chemistry and allied subjects, 1900-1924* (Washington, National Research Council, 1925) and *Supplements, 1925-31* (1929-32); R Lauche *World bibliography of agricultural bibliographies* (Munich, BLV, 1964). He will also be aware of the difficulties of locating the so-called ' hidden ' bibliographies, *ie*, those that appear as articles in journals, or as part of a monograph, or are otherwise not separately published. He will probably know that the National Reference Library of Science and Invention maintains a subject index on cards of bibliographies, including ' hidden ' bibliographies. An even greater problem is bibliographies that are not published at all: a very large number of these exist in universities, research institutes, libraries, etc, typically in the form of a file of references on cards or slips. As often as not they are freely available for consultation, and would indeed be published, if the time and effort could be spared. Instances of files of this kind that have been published are the Boyce Thompson Institute for Plant Research, Yonkers, NY, collection of 30,000 references built up over 20 years which appeared as L V Barton *Bibliography of seeds* (Columbia UP, 1967); and the Plutonium Kartei in Karlsruhe, a comprehensive bibliography on 11,000 cards on the element plutonium, which from January 1967 was converted into a monthly abstract journal, *Plutonium-Dokumentation*. And like library catalogues, such files can now be comparatively effortlessly reproduced by photolitho-offset, *eg*, the Gray Herbarium Index to new plants since 1886 on 265,000 cards at Harvard University, published in ten volumes (Boston, Hall, 1968).

It will be evident to the student reading this that he has stumbled into an area where bibliographical control is deficient. It may surprise him to learn that there is no fool-proof way of tracing a bibliography on a particular subject without actually carrying out an exhaustive search. Even the most knowledgeable librarian cannot guarantee that the relevant bibliographies will be encountered at the beginning rather than during the course of a literature search. There is some hope here that can be glimpsed in the greatly increased manipulative power that computers can offer: it is comparatively simple at the input stage to ensure that citations consisting of or containing bibliographies are 'tagged' as such. Then at the retrieval stage an appropriate search strategy will see to it that such references emerge first. On the other hand, mechanised systems (like MEDLARS and *Chemical abstracts*) are adding to the problem of control with the facilities they offer for on-demand, custom-produced bibliographies.

That there is need for coordination of bibliographical effort is evidenced by flagrant instances of duplication, *eg*, J H Batchelor *Operations research: an annotated bibliography* (St Louis UP, second edition 1959), and Case Institute of Technology *A comprehensive bibliography on operations research* (New York, Wiley, 1958), both supplemented by further volumes; and V E Coslett *Bibliography of electron microscopy* (Longmans, 1951) covering 1927-48, and C Marton *Bibliography of electron microscopy* (Washington, USGPO, 1950) covering 1926-49. The compilation of bibliographies is an area eminently suited to cooperative endeavour: by its nature it lends itself to precise divisions of labour and of responsibility and even the smallest contributions can play their significant part in a total scheme. The most obvious manifestation of such cooperation is the union list, *eg*, American Chemical Society Rubber Division Library *Union list of books relating to the fields of rubber, resins, plastics and textiles held by* [eight] *technical libraries* ... (Akron, Ohio, 1962), and Cement and Concrete Association *Bibliography of cement and concrete: list of books and papers in London libraries* (third edition 1952). Other instances are: K M Clayton *A bibliography of geomorphology* (Philip, 1964) which covers books and articles 1945-62, and is deliberately designed to dovetail with *Geomorphological abstracts*, which commenced in 1963; W H Mullens *A geographical bibliography of British ornithology, from the earliest times to the end of 1918* (Wheldon and Wesley, 1919-20), the arrangement of which under counties neatly complements W H Mullens and H K Swann *A bibliography of British ornithology from the earliest times to the end of 1912* (Macmillan, 1916-17) and *Supplement* (1923), which is arranged alphabetically by author; and the US National Advisory Committee for Aeronautics *Bibliography of aeronautics, 1909-1932* (Washington, USGPO, 1921-36), which in fourteen volumes and following the same basic plan serves as a continuation to the remarkable Paul Brockett *Bibliography of aeronautics* (Washington, Smithsonian Institution, 1910), which revealed that by July 1909 there were no fewer than 13,500 titles extant on this comparatively novel subject.

As has been mentioned, the range of subjects that do have their own bibliographies is quite extraordinary. Even more impressive in many cases is the extent of the published literature on quite specialised topics that these bibliographies unfold: L M Cross *A bibliography of electronic music* (Toronto UP, 1967) lists 1,562 writings in five languages; the computer-produced *Laser literature, 1958-1966* (North Hollywood, Calif, Western Periodicals, 1968) in two volumes has 4,357 entries; John Greenway *Bibliography of the Australian aborigines* (Angus and Robertson, 1963) has over 10,000 references; W F Clapp and Roman Kenk *Marine borers: an annotated bibliography* (Washington, Office of Naval Research, 1963) takes 1,148 pages to list the world's literature on ship-worms.

As the student will be aware, the art of compiling and arranging bibliographies, and a consideration of the variety of form, content, and coverage of published examples is a study in itself, beyond the scope of this chapter. Nevertheless, there still remain a number of distinct categories of bibliography commonly encountered in science and technology that are deserving of particular attention. One of the most interesting is the ' narrative ' bibliography, where the citations are dispersed through a continuous explanatory text. The best of these can provide a useful introduction to the subject as well as a list of books, but the added comment sometimes brings them close to the guides to the literature discussed in chapter 2, *eg*, J L Thornton and R I J Tully *Scientific books, libraries and collectors* (Library Association, second edition 1962). Even more than the unadorned enumerative bibliographies, such works benefit greatly where the compiler is an authority on his subject as well as a bibliographer: the result can often turn out to be a ' tour de force ', as in the case of E T Bryant *Railways: a readers guide* (Bingley, 1968).

A fruitful source of information are those bibliographies (usually of very specific subjects) which are issued by libraries as items in a series. Perhaps the best known is the Science Library Bibliographical Series which has been running since 1931 and is now close to its 800th title: examples of topics covered are semiconductors, manufacture of glass mirrors, papermaking. Since most of such lists are compiled originally in response to a request, it is obvious that they hold considerable current interest. They are often available from the issuing library merely for the asking, they are almost invariably based on extensive subject collections and frequently they are compiled by subject experts. Well established series are the (us) National Agricultural Library Lists (no 95, published in 1969, is *Sunflower: a literature survey, January 1960-June 1967*, with over 1,800 references); the Atomic Energy Research Establishment (Harwell) Bibliographies (no 164 is R Morgan *A selected bibliographical guide to conference papers on nondestructive testing, 1955-67*, (1968); the Iron and Steel Institute Bibliographical Series (no 24 (1965) comprises 600 references on *Vacuum metallurgy of steel*).

Similar in some ways to the library catalogue insofar as it is not primarily compiled as a bibliography is the exhibition catalogue. From time to time special displays of books are gathered together for a limited time, sometimes to mark a particular occasion such as a centenary, and then they are dispersed, perhaps for ever. If the exhibition is really important, its catalogue may have permanent value as a subject bibliography, *eg*, Science Museum *A hundred alchemical books* (HMSO, 1952); H D Horblitt *One hundred books famous in science: based on an exhibition held at the Grolier Club* (New York, 1964); National Book League *Do-it-yourself: a touring exhibition* (1962). Catalogues of private collections too can serve as valuable additions to the bibliography in their field, where the collection is sufficiently worthy, *eg*, *Catalogue of botanical books in the collection*

of *Rachel McMasters Miller Hunt* (Pittsburgh, 1958-61), two volumes
in three; *Bibliotheca alchemica and chemica: an annotated catalogue
of printed books . . . in the library of Denis I Duveen* (Weil, 1949).
Regrettably, such catalogues are sometimes compiled as a preliminary
to the sale or dispersal of the collections, but John Ferguson *Bibliotheca
chemica: a catalogue of the alchemical, chemical and pharmaceutical
books in the collection of the late James Young* (Glasgow, Maclehose,
1906), in two volumes, and *Bibliotheca Osleriana: a catalogue of books
illustrating the history of medicine and science, collected . . . by Sir
William Osler* (Oxford, Clarendon Press, 1929), both describe collections that were bequeathed as a whole (to Strathclyde and McGill
universities respectively). Of similar bibliographical interest are some
of the sale catalogues of antiquarian booksellers: an interesting bibliography compiled from just such sale catalogues of a London book-
seller is H Zeitlinger and H C Sotheran *Bibliotheca chemico-mathe-
matica: catalogue of works in many tongues on exact and applied
science* (Sotheran, 1921) in two volumes with *Supplements* (1932,
1937, 1952).

Obviously, catalogues of exhibitions and of private collections tend
to be of mainly historical interest, and the same is probably true of
most author bibliographies in the field of science and technology. Of
course they are not *subject* bibliographies at all, strictly speaking,
but they can serve as a further source to consult on occasion. They
fall into two main categories: those which confine themselves to listing
the writings by a particular scientist or technologist, and those which
in addition list writings *about* him. Examples of the first kind are
Nell Boni and others *A bibliographical checklist and index to the
published writings of Albert Einstein* (Paterson, NJ, Pageant Books,
1960) and A E Jeffreys *Michael Faraday: a list of his lectures and
public writings* (Chapman and Hall, 1960). Examples of bibliographies with works by and about an author are J F Fulton *A bibliography of the Honourable Robert Boyle* (Oxford, Clarendon Press,
1961), A L Smyth *John Dalton, 1766-1844: a bibliography of works
by and about him* (Manchester UP, 1966), and R E Crook *A bibliography of Joseph Priestley, 1733-1804* (Library Association, 1966).

A feature of this last title is the location of copies of listed works
in over two hundred libraries in a dozen or so countries. This reminds
us that for practical purposes the tracing of a wanted item in a bibliography is only half the battle, and compilers who take the trouble to

indicate where a copy can be consulted (or borrowed, or photocopied, or microfilmed) perform a valued service, *eg*, R M Strong *A bibliography of birds* (Chicago, Natural History Museum, 1939-46), with 25,000 entries in three volumes.

9

PERIODICALS

A well-nigh universal finding of the many surveys of the literature habits of scientists and technologists is that the most frequently used of all sources of information are periodicals. Known variously as journals, serials, magazines, transactions, proceedings, bulletins, etc, and commonly appearing weekly, monthly or quarterly, their value cannot be over-estimated. They form the heart of most specialist collections and in many scientific and technological libraries more is spent on periodicals than on books. Furthermore, the content of most books in science and technology is based on the periodical literature of previous years.

It is generally agreed that modern science had its beginnings in the seventeenth century: it is not without significance that the invention of the periodical dates from that time also. Although the printed book had been serving as an increasingly important medium of communication for two hundred years, the developing community of scientists found it necessary to supplement its inadequacies in 1665 by founding the *Philosophical transactions of the Royal Society,* the earliest learned journal in any subject still in existence. The book, it was felt, was more suited to the formal and proper publication of mature reflections or a completed *opus,* and was not for reporting current discoveries. This new medium, on the other hand, was specifically designed to print research material of this kind, which scientists until that time had been announcing in personal correspondence or at meetings with their colleagues. This intimate connection between scientific discovery and the scientific periodical has continued until the present day, and the exponential growth of science is paralleled by a similar increase in the number of scientific periodicals. By 1800 there were about 100 titles, and by 1900 it is estimated that this figure had reached 5,000. The most reliable recent estimates indicate that in science and technology the number of current periodicals is approaching 30,000.[1]

[1] K P Barr 'Estimates of the number of currently available scientific and technical periodicals' *Journal of documentation* 23 1967 110-6.

The continued fragmentation of science is another noteworthy feature that can be matched in the periodicals. The early journals, like early science, were undifferentiated, embracing the whole of ' natural philosophy ', as it was then called, and much else. Indeed at that time even the boundary between science and technology was very often ignored: the charters of the Royal Society (1662 and 1663) specifically refer to 'further promoting by the authority of experiments the sciences of natural things *and of useful arts*'. But as we know, ' general science ' was soon supplemented by the individual disciplines of chemistry, biology, physics, which in turn developed divisions such as pure and applied chemistry, and then further aspects like organic and inorganic chemistry. Sub-division still continues apace, as the price we pay for increasing knowledge, and now to the layman seemingly minute specialisations like chromatography or carbohydrate chemistry are established disciplines in their own right. It is an illuminating exercise for the student to trace these developments as reflected in the establishment of new journals in a discipline over a period of years.

This fragmentation of science is demonstrated even more dramatically by the increasing number of journals showing fissiparous tendencies: in 1966 the *Journal of the Chemical Society* split into three parts after more than a century as the major periodical in its field; more recently the *Proceedings of the Physical Society* have similarly divided into the tripartite *Journal of physics;* the *American Society of Mechanical Engineers Transactions* and the *Institute of Electrical and Electronic Engineers Transactions* took a similar course some years ago. There are many others.

All this is not to imply that periodicals covering the whole field of science are now superfluous. On the contrary, they are needed more than ever to play two specific roles: firstly, to act as a counter to the centrifugal tendencies of modern science by providing a forum for different kinds of specialists and attempting co-ordination across disciplinary boundaries, *eg, Nature, Philosophical journal, Proceedings of the Royal Institution of Great Britain;* and secondly, to give scientists an opportunity to explain, particularly to non-scientists but also to specialists in other fields, the implications of their work, *eg Science, New scientist, Scientific American, Science journal,* each of which, to quote from the policy statement of one of the best examples of this category of journal, *The advancement of science,* are ' of interest to laymen wishing to understand something of the significance of science and its impact on society '.

THE IMPORTANCE OF PERIODICALS
Before proceeding further it is essential for the student to understand why periodicals hold so pre-eminent a place in the literature of science and technology, and a good way to start is to examine the ways they differ from books. R L Collison has written: ' The extent of knowledge in any field consists . . . of the information given in the books on the subject *plus* the periodical articles that have been published since the latest book was written '. In science and technology at least, no worker would contemplate writing up his *latest* research in the form of a book. In the first place, the simple speed of printing and distribution of the periodical compared with the book, deriving from the comparative brevity of the papers, the guaranteed (usually pre-paid circulation), and the unbound format, means that it is usually months and often years more up-to-date than its rival; secondly, the regular and frequent appearance of the periodical ensures that it can *remain* close to the frontiers of knowledge by continuous addition, correction, or even retraction. In the words of E J Crane *A guide to the literature of chemistry*, 'A book is soon out of date, but a live journal can and does keep up with the onward march of scientific discovery '. These are the features of the periodical which have also enabled it over the years to perform the social rather than truly scientific function of establishing priority for the work of the individual scientist.

The average citizen is often astonished to learn that there are considerable numbers of topics on which no information has been published in book form. It is true that many are new subjects which will appear in the books in due course, but that still leaves thousands which are too brief, or insignificant, or local, or ephemeral, or recondite, ever to warrant full-scale treatment in books. For topics such as this (as for new subjects) the only source of printed information may well be an article in a periodical. Furthermore, even where books are available on a particular subject, periodical articles can often furnish more detailed (not to say accurate) information on background, history, methods, apparatus, etc. After all, it is the periodical that is the primary source: the book is secondary, based largely, if not entirely, on the periodical literature.

By their nature, periodicals perform a further range of very useful tasks impossible for books. They contain correspondence columns often of great importance, permitting exchange of views about papers published (or any relevant topic). Many carry book reviews, obituaries, editorial comment, abstracts. Current news, professional announcements, advertisements, are also commonly found. In many cases they serve as ' organs ', often of a society or group, but in some cases merely of that band of like-minded individuals making up the subscribers, much in the same way as does the *Daily express* newspaper, which once described its readers as members of one large, happy family.

Paradoxically perhaps, in view of some of the difficulties which will be outlined later, the major periodicals in science and technology are often easier of access than books; that is to say they are better indexed, abstracted, and located. Because of their prime importance in science and technology the bibliographical care normally exercised on books has been devoted to the production of a remarkable range of bibliographies, indexing and abstracting services, and location tools.

CATEGORIES OF PERIODICALS

Faced with the tens of thousands of current periodical titles in science and technology, the student could well be excused his sigh of despair. And it must be confessed that any attempt at classification by type can only be arbitrary and at times conflicting, although for purposes of study it must be attempted.

There is no dispute, however, about the basic division into primary and secondary journals. The primary journals, of course, devote themselves to reporting original research. Known also as ' recording ' journals, they form the bedrock of scientific and technological literature, *eg, Biochemical journal, Journal of physiology, Journal of mechanical engineering science, Journal of the Institution of Electrical Engineers, Computer journal, Tetrahedron, Philosophical magazine, Transactions of the Faraday Society, Molecular physics.* The task of the secondary journals, on the other hand, is to digest, comment on, and interpret the research reported in the primary literature. They have been called ' newspaper ' journals, but they make up a far more heterogeneous collection than the research journals, *eg, Chemistry in Britain, Glass, Engineering, Colliery guardian, American machinist, Textile industries, Physics today, Chemical and engineering news, Plumbing trade journal, Production technology.* And of course there are many hybrids, containing both research papers and secondary material, *eg, Spaceflight, Production engineer, Journal of the Electrochemical Society, Chemistry and industry.*

There is, however, a particular manifestation of the secondary journal that has been the object of such close attention in recent years and is obviously destined to play so crucial a role in scientific and technological communication in the future that it is worth elevating into a special third category. This is the ' review ' journal, comprising articles that briefly survey developments in a particular field of endeavour over a period, eg, *Biological reviews, Advances in physics, Metallurgical reviews, Science progress.* The review journal will be considered at length with other kinds of reviews of progress in chapter 11.

One commonly used *ad hoc* classification of periodicals for purposes of study is by their origin,[2] as follows:

a) *Learned societies, academic bodies*: as we have seen, the periodical as a form of literature owes its origin to the learned society, and since the earliest days such societies have been responsible for a significant proportion of the total, eg, *Journal of the Chemical Society, Journal of heredity* (American Genetic Association), *Journal of physical chemistry* (American Chemical Society), *Proceedings of the National Academy of Sciences* (USA), *Proceedings of the Cambridge Philosophical Society* have all been established for over fifty years. The main purpose of such periodicals is to furnish an opportunity for authors (usually members of the learned bodies concerned) to publish the results of their investigations, and perhaps the majority of titles in this group are research journals, but there are also a number of secondary journals issued by the societies, frequently alongside a primary journal; eg, in addition to its research quarterly *Computer journal* the British Computer Society also brings out the monthly *Computer bulletin* as its ' organ ', with reports of meetings, data on new equipment, additions to the library, etc. As well as its century-old *Proceedings* the Cambridge Philosophical Society produces the quarterly *Biological reviews*, surveying four topics of current interest in each issue. A large society like the American Mathematical Society finds it needs *two* research journals (*Proceedings* and *Transactions*) and two others (*Bulletin* and *Notices*). Nevertheless, the main contribution of these societies to the literature is their still very substantial role in dissemin-

---

[2] In most cases, perhaps, this is the same as classification by *publisher*. But not always: *Clio medica* is the official journal of the International Academy of the History of Medicine: *Solar energy* is a journal of Amazon State University; both are *published* by Pergamon Press, a commercial publishing house.

ating the reports of original research through their splendid scholarly journals such as *Applied optics* (Optical Society of America), *Journal of physics* (Institute of Physics and Physical Society), *Journal of applied chemistry* (Society of Chemical Industry), *Journal of organic chemistry* (American Chemical Society).

It is a commonplace of science today that research has become institutionalised, and that the lone scientist is now a very rare bird. Since much of this new research is undertaken in academic institutions it is perhaps not surprising to find it increasingly reported in university and college research journals, many of which have been established for a number of years. As yet they have not the prestige or circulation of the great learned society titles, but they are indicative of a trend, *eg, Japanese journal of physics* (University of Tokyo), *Bulletin of mechanical engineering education* (University of Manchester Institute of Science and Technology), *Journal of geology* (University of Chicago), *Annals of tropical medicine and parasitology* (University of Liverpool).

b) *Governmental bodies*: as the role played by government, both national and international, in our lives increases, so does the volume of official publication, particularly in science and technology (see page 170), where vast sums of public money are currently being spent on research and development. Some of these publications are periodicals, *eg, Meteorological magazine, Marine observer* (both Meteorological Office journals), *Post Office telecommunications journal, Plant pathology* (Ministry of Agriculture), *Journal of research of the National Bureau of Standards* (USA), *Australian journal of biological science* (Commonwealth Scientific and Industrial Research Organisation), *Canadian journal of chemistry* (National Research Council of Canada), *World health* (World Health Organisation).

c) *Independent research institutes*: a small but interesting group of periodicals emanates from research institutes that are basically of independent foundation (even though they perhaps have links with universities, or possibly undertake government work under contract). They may have been established with a particular subject orientation or a particular role to play, or they may be ' think-tanks ' on the American pattern, willing to undertake research in a variety of disciplines on a bespoke basis. Examples of periodicals so produced are *Battelle technical review* (Battelle Memorial Institute, Columbus, Ohio), *Textile research journal* (Textile Research Institute, Princeton, NJ), and *Polar record* (Scott Polar Research Institute, Cambridge).

d) *Professional bodies*: as a category, bodies like the Institution of Mechanical Engineers, the Royal Institute of Chemistry, the Royal Institute of British Architects, overlap with the learned societies, and much of their work (and the periodicals they produce) is indistinguishable. Nevertheless, it is possible (and useful) to consider them separately as disseminators of scientific and technological literature. In broad terms (and at the risk of oversimplifying) it can be said that the concern of the learned society is the promotion of its subject, whether biology or physics or mathematics; in the professional body consideration of the education, welfare, status, etc, of the practitioners is added to a concern for the subject.

Periodicals in this category can range from primary research journals of a calibre and prestige fit to match any learned society publication to what are little more than news bulletins. And the fact that the scientific and technological professions vary so much in their size, cohesiveness, sense of professional responsibility, and status leads to an even wider range of publications. Examples to study and compare are *Mathematical gazette* (Mathematical Association: ' an association of teachers and students in elementary mathematics '), *Plastics and polymers* (Plastics Institute), *Structural engineer* (Institute of Structural Engineers), *Quarry managers' journal, Post Office electrical engineers' journal, Concrete* (Concrete Society), *Journal of basic engineering* (American Society of Mechanical Engineers), *Chartered mechanical engineer* (Institution of Mechanical Engineers). And like the learned societies, some of these professional institutions find they need more than one journal to serve their needs. An instructive instance is the long-established and prestigious Textile Institute, with its three regular publications neatly illustrating the three categories of periodicals described at the beginning of this section: *Journal of the Textile Institute*, for the publication of original research work and detailed accounts of practical investigations; *Textile Institute and industry*, with briefer papers of a more practical bias, ' especially acceptable to the busy executive '; and *Textile progress*, each issue of which is devoted to a full-length critical review of a major topic of technological interest, *eg*, textile machinery.

e) *Commercial publishers*: the overwhelming majority of periodicals in science and technology are produced as a business venture by commercial publishing houses.[3] Since they must make a profit to

---

[3] This statement is not true of communist countries, of course.

survive, the titles are understandably heavily concentrated at the applied, industrial and commercial, technical and trade end of the spectrum, *eg, Paint technology, Traffic engineering and control, Commercial motor, Water and water engineering, Welding and metal fabrication, Rubber journal, Electrical and radio trading.* But this group contains all types and levels of periodical publication and even of the most significant scientific journals the commercial publishers produce a substantial share: of the top 165 British titles (ranked according to the number of citations they receive) over half are commercial publications and a further 26 are society or professional journals published by a commercial publisher.[4]

So wide is the variety of such periodicals that it is expedient to subdivide them further into:

(i) *learned and research periodicals*: examples have been commercially produced for a hundred years or more, particularly in Germany, but until recently they have always been overshadowed by the famous titles issued by the learned and professional societies. Over the last two decades it seems that these societies have lacked the financial resources (and possibly the commercial enterprise) to initiate further titles, despite the vast increase in the amount of research work in need of a literary outlet. Into the breach have stepped scientific and technological publishing houses such as Elsevier, Interscience, Taylor and Francis, Academic Press, Pergamon, and we now have the phenomenon of several hundred primary research journals, a significant sector of the pattern of scientific communication, appearing apparently successfully as a commercial undertaking. Very considerable efforts have been made to ensure that the highest standards are maintained, and it is usual for each such journal to have an editorial board of eminent scientists, insulated so far as is possible from commercial pressures. While some commercial journals contain too large a proportion of inferior papers, there is no doubt that others have reported work of the highest quality. Representative titles are *Journal of molecular biology, Talanta, Polymer, Journal of chromatography, Journal of the mechanics and physics of solids, Annals of physics, Microchemical journal.* This development has coincided with a huge absolute and relative increase in institutional (mainly library) buying, and

[4] John Martyn and Alan Gilchrist *An evaluation of British scientific journals* (ASLIB, 1968).

unfortunately there is evidence that this captive market is being exploited by exorbitant prices.[5]

(ii) *technical journals*: these are very closely linked with the needs of industry, and although as secondary sources are of limited interest to the research investigator, they are invaluable to manufacturing, sales, and commercial personnel. Their role is to digest, interpret and comment as well as to inform, and their level is that of the practising chemist, or working engineer, or technically-minded manager: commonly they contain generously illustrated papers on new processes, equipment, products and materials. The fact that they are less academic than the research journals, and do not perhaps function at such a high level of scientific activity, does not mean that the papers they print are not often very technical, advanced and scholarly. Indeed, it is possible to draw too firm a line between some titles in this group and the primary research journals. Let the student examine, for instance, *Ultrasonics* or *Chemical engineering*.

Much of their current value lies in their other features, such as news columns, letters to the editor, announcements, obituaries and other personalia, book reviews, lists and abstracts of other literature, especially patents and trade literature, etc. Particularly useful are the advertisements, of which there are usually a great number.[6] Most of these are deliberately designed to inform, making them totally different from consumer advertisements. It is quite common to find in such advertisements for new machinery, or chemicals, or processes, for instance, details unavailable elsewhere in the literature. The index to advertisers frequently provided in each issue of a journal is but one indication of their value. The buyers' guides and directories sometimes issued as supplements (see page 64) are a further well-appreciated

[5] It is interesting to note where some would place the responsibility for this. One scientist writes: ' It is idle to blame the publisher for making as large a profit as he can: it is his business to do so; but it is equally the business of the scientist to prevent excesses. The captive librarian is helpless in this situation, for he must buy what his readers require and is not free to renounce any journal of sound content; it is the scientist alone who carries the responsibility, and it is his duty to prevent extortion and exploitation; if high-ranking editors will refuse to collect papers, or senior authors to submit them, then the price of the journal will inevitably be reduced to the " fair profit " level.' R S Cahn *Survey of chemical publications* (Chemical Society, 1965), 29.

[6] Advertisement revenue is the life-blood of such journals, as it is of the mass media, and there is in some quarters the same concern that it should not divert the technical journal from its main function. There are signs that some ' technological glossies ' have allowed this to happen.

service, *eg*, ' *Soap, perfumery and cosmetics* ' *year book and buyers' guide.*

Their retrospective importance in the literature of science and technology is obviously less than their current value, and depends on the proportion of material with some claim to permanence that appears within their covers. Examples of titles to study are *Electronic engineering, Foundry trade journal, Food engineering, American dyestuff reporter, Computers and automation, Engineer, Wood, Vacuum, Wire industry, Surveyor, Shipping world and shipbuilder.*

(iii) *trade journals*: between these and the previous group the dividing line is vague, with a great deal of overlap and many cross-bred examples, but in general they are more commercial than technical, and more news-oriented than subject-oriented. Otherwise they are very similar to the technical journals, with an equal reliance on advertisements. Many cover a small well-defined sector, and a substantial number were founded in the nineteenth century, *eg*, *Poultry world* (1874), *Paint, oil and colour journal* (1879), *Bakers' review* (1887), *Printing world* (1878), *Veterinary record* (1888), *Commercial grower* (1895), *Contract journal* (1879). For some specialised topics, the file of the appropriate trade journal may form the greater part of the total published literature on the subject.

Such journals are particularly useful sources for market news (commodity and share prices), company news (forecasts, dividends, mergers, expansions), and general trade announcements. A large number appear weekly, *eg*, *Wool record, Shoe and leather news, Machinery market,* and *Mining journal,* as well as the seven titles mentioned in the previous paragraph, and are often officially registered as newspapers. Some have even adopted newspaper (tabloid) format, *eg*, *Construction news, Electronics weekly, Farmers guardian.* They are surprisingly little known outside their respective trades, and deserve far more attention from libraries and a much wider readership among the scientific and technological community than they get. Excellent examples for the student to examine are *Chemical age, Motor industry, Farmer and stockbreeder, Timber trades journal, Sewing machine times, Textile and woollen trades gazette.*

In his exploration of technical and trade journals the student is likely to come across a number of ' controlled-circulation ' periodicals, such as *Instrument and apparatus news, Electrical equipment, Industrial equipment news, Maintenance engineering, Food engineer, Process engineering, Transport journal, Scientific research.* These are almost entirely advertisement media, and are usually sent free of charge to those considered to be qualified readers, *ie*, actual prospective customers for the goods advertised, or at least those in a position to influence purchasing decisions. Some are available on request to libraries as well as to individuals, and they do have some value as sources of otherwise unavailable data.

(iv) *popular subject journals*: these are familiar to everyone, and include all the titles for the amateur, the hobbyist, and the enthusiast that are to be found on railway bookstalls, as well as a large number of esoteric (not to say crank) publications catering for the most unusual preoccupations. Typical titles are *Popular mechanics, Yachting world, Flying saucer review, Railway magazine, Practical motoring, Amateur gardening, Radio constructor, Racing pigeon, Model engineer, Speleologist, Inventor*. Again it should be pointed out that this is not a watertight group. Serving the needs of the amateur is a trade for some, and for certain areas of interest it can be an industry. Periodicals in the field often double as trade or technical journals and popular journals, *eg, Tape recorder, Motor boat and yachting, Bee craft, Entomologists' record, Microscope*. There are instances of subjects like woodworking or horsebreeding or horticulture where many amateurs are as knowledgeable as the professionals, and their journals reflect this. *Wireless world* is often quoted as an example of an originally popular magazine that is now a highly technical journal. And those inclined to scoff at the lunatic fringe should study what has happened to the status of the *Journal of British Interplanetary Society* since its establishment in 1934.

e) *Industrial and commercial firms*: there are in English perhaps over ten thousand ' house journals ' or ' house organs ', issued primarily for advertising purposes by manufacturers and dealers and public corporations, *eg, Educational focus* (Bausch & Lomb Optical Co), *Laboratory* (Fisher Scientific Co), *BOAC review, Ball bearing journal* (Skefco Ball Bearing Co), *Philips technical review, Midland Silicone news, Hexagon Courier* (ICI Ltd, Dyestuffs Division). Like other forms of trade literature, however, they often contain information of value which may be unobtainable elsewhere. They are described in more detail in chapter 17 (pages 201-3).

f) *Individuals*: these are now quite rare, although a number of today's most important journals commenced as the production of a single man. They were particularly prevalent in Germany, and perhaps the best known instance (which still retains the founder's name in the title) is *Justus Liebigs Annalen der Chemie*.

THE PROBLEMS OF PERIODICALS

Of the importance of periodicals to the scientist and technologist no one has any doubts, but as R L Collison has pointed out, ' . . . the full exploitation of periodicals is one of the most important problems which faces the librarian in any type of library'. As we have seen, the periodical owes its origin to the seventeenth century scientists' dissatisfaction with the printed book as a medium of communication, but it never provided the complete answer, and scientists are still far from content with their literature. The periodical in particular has come in for severe criticism in recent years.[7]

In the first place there are far too many: to the so called ' information explosion ' periodicals have contributed as much as or probably more than any other category of non-book materials. Over twenty years ago Sir Edward Appleton, Nobel prize-winning physicist and discoverer of the ionosphere, wrote: ' If anyone set himself the task of merely reading—let alone trying to understand—all the journals of fundamental science published, and worked solidly at his task every day for a year, he would discover at the end of the year that he was already more than 10 years behind! If the same constant reader had included the technical literature as well, he would find himself about 100 years behind in his work after 12 months' effort.' Since then, of course, the number of journals has continued to increase, according to the estimates doubling in total every ten or fifteen years.

Paradoxically, as the number grows, evidence is accumulating that actual use is becoming relatively more restricted. By using the number of times an article is cited as a rough-and-ready measure of its importance, studies invariably show a large majority of the articles concentrated in a minority of journals, *eg*, in computer literature 75 percent of references are to be found in 17·6 percent of the journals; 84·3 percent of the references in 26 volumes of *Physical review* are to 15

---

[7] *eg*, ' Despite the invention of photography and microfilms and television and computers, we cling to the publishing habits of our forefathers. Though the amount of material to be communicated is fabulously greater, a very large proportion is transmitted through journals which . . . are essentially the same as they were a century or more ago. And where one journal does not suffice, we seldom think of anything cleverer than to create another like it; and another.' Sir Theodore Fox *Crisis in communication: the functions and future of medical journals* (Athlone Press, 1965).

journals, less than 2 percent of the total number of journals referred to. The most extensive recent survey, by Martyn and Gilchrist for ASLIB (see page 106), investigated references to British journals in a sample year's issues of *Science citation index,* and discovered that 95 percent of the 68,764 citations relate to a mere 9 percent (165 titles) out of a possible 1,842 British scientific journals. With D J de Solla Price, the student is 'tempted to conclude that a very large fraction of the alleged 35,000 journals now current must be reckoned as merely a distant background noise, and as very far from central or strategic in any of the knitted strips from which the fabric of science is woven'. Research results of this kind have suggested the concept, of great interest to librarians, of a collection of 'core' journals, limited in number, but sufficient to meet perhaps 95 percent of the demand. The ASLIB investigators conclude that 'a rough estimate of the world's number of core science journals lies between 2,300 and 3,200'. Studies of the use of scientific and technological periodicals in libraries tend to support this. One of the largest ever undertaken, covering 87,255 issues from the Science Museum Library in 1956, showed that 'Even in a library which is designed to deal with the residual demand from libraries, about 1,250 serials (or less than 10 percent of those available if the non-current serials are included) are sufficient to meet 80 percent of the demand for serial literature'. Surveys of scientists' and technologists' literature habits indicate that the core journals in any specialist discipline may not amount to more than half-a-dozen: the Wood and Hamilton study of mechanical engineers noted below (page 149) found that only 12 percent see more than ten journals regularly and 37 percent see less than five; and an earlier survey in the electrical and electronics industries discovered that the average was 4·7. Yet despite the vast numbers of periodicals, and their obvious spare capacity for good articles, coverage of the field seems to be inadequate. The complaint is frequently heard that much work of significance is not published for lack of space, and new journals appear constantly.

A common cause of grievance for the authors is the delay between submission of a paper to a journal and its publication, which is quite often six months and sometimes more. To some extent the cure is in the authors' own hands, for if they will insist on concentrating their attentions on the better-known core journals an inevitable queue must result. Part of the delay is caused at the editorial checkpoint, for it is common practice with learned journals in English-speaking countries to send submitted papers to an independent 'referee' for an authoritative opinion before publication. This certainly helps to preserve standards by eliminating the occasional worthless or even fraudulent paper, but authors naturally find it burdensome, and even the most august referees are not infallible: Sir Theodore Fox tells us that one of the discoverers of blood-groups revenged himself by keeping the *Lancet*'s letter of rejection framed on the wall of his consulting-room. Delay is particularly frustrating in science and technology where the pace of advance is so great: the very pressures toward rapid publication which created the periodical in the first place are now being brought to bear on the periodical itself.

One of the most telling arguments against the periodical is its wastefulness. A UNESCO study has shown that a single article in a highly specialised periodical may interest no more than 10 percent of the workers in the subject area of that periodical, while a single article in a general periodical will interest only about 2 percent of its readers. J D Bernal has said that in writing papers for journals instead of personal letters, scientists have replaced communication by dissemination, *ie*, by broadcast scattering. One is bound to suspect that many articles are not read by anybody, but are published just for the record, often merely to establish a worker's priority in the field. As H V Wyatt has written, ' For the individual, the modern scientific paper is a social device for proclaiming intellectual property, not primarily a technique for advancing science '. A *Nature* editorial is more pithy: ' . . . the habit of writing for posterity is often an impediment to communication with those still alive '. Some waste is perhaps inevitable in the process of communication, but there are now several essential journals costing more than £50 ($120) a year, while cost-benefit analysis stalks the land. Economic necessity has obliged a number of US learned and professional societies to institute ' page charges ' payable by the contributors to their journals in an effort to keep subscription prices down. Many workers are indeed beginning to agree with the opinion that ' . . . the entire apparatus of commercial publication is too expensive a mechanism for scholar-to-scholar communication '.

As the student knows, science and technology by its nature is progressive and cumulative, and inevitably much of its literature is ephemeral. But the time when any given article is going to be superseded cannot be predicted, and even the average rate of obsolescence differs greatly according to subject. As a general rule of thumb, every year perhaps 10 percent of scientific and technological papers ' die ', never to be cited again, but this rule does not apply to a number of important disciplines where classic papers may retain their importance for generations, *eg*, botany, geology, mathematics. Many chemists bank on the significant journal literature becoming incorporated in the book literature within twenty years, but this would certainly not hold good for biologists. The massive survey of Science Museum Library issues in 1956 (see page 111) showed that journals published within the previous seven years accounted for 57 percent of the use, while journals over 27 years old made up less than 9 percent of the issues. Considerations of this kind are of vital concern to the librarian, who has to decide, often in the face of conflicting demands for space, how long to retain back runs of journals.

Bibliographical control of periodicals and their contents will be discussed later in this chapter and the next, but it is appropriate here to look at one of the major obstacles to adequate control, and therefore to satisfactory use. This is the interesting phenomenon of ' scattering ', *ie,* the extent to which articles on a given subject actually occur in periodicals devoted to quite other subjects. This has been observed for very many years, but was first studied in detail by S C Bradford of the Science Museum Library, who codified his findings in what has become known as Bradford's Law of Scattering. His work has been refined and added to by a number of investigators since, but in very crude and over-simplified terms the Law states that of all the articles on a subject, one third are to be found in journals on that subject, one third in journals on related subjects and one third in quite unexpected journals. The implications for a scientist or technologist making an exhaustive approach (see page 18) to a subject are obvious: he cannot hope to locate all the required information without ranging widely outside the journals in his own field. One might add here that a problem of a similar nature is sometimes encountered with regard to the quality of research papers: the well-known journals can ensure that a poor paper is never published in their pages, but what no one can prevent is a paper of the highest class appearing in a lowly and obscure journal.

A word too should be said here on the bibliographical idiosyncrasies of periodicals: the student should school himself never to show surprise at any aberration of titling, numbering, dating, frequency, format, pagination, or publisher. The first issue of *Traffic engineering and control,* for instance, was number 2; volume 40 (1966) of *US government research and development reports* was followed immediately by volume 67 (1967); the index of *Automotive design and engineering* is two inches taller than the journal itself; in 1957 the periodical known for nearly twenty years as the *Muck shifter* changed its title to *Public works*, but in 1962 it changed back again; there is one periodical which changed its name 41 times in 14 years. And one should not forget that the life span of the *average* periodical is less than ten years.

Periodicals are subject to a bibliographical hazard that most books escape—the very common practice of abbreviating their titles in citations, *eg*, J Chem Soc, Engr, Paper Tr J, Mfg Chem, Sci Prog, Phys Rev. Not all are as self-evident as these, *eg*, Text Rec, JACS, ASME Proc, and one still encounters unofficial, home-made, inconsistent abbreviations. Unfortunately, there are at least half-a-dozen official systems, all different but still widely used. Sterling efforts are being made towards standardisation, but when it comes it cannot alter the millions of confusing abbreviations in the literature to date.

ALTERNATIVES TO THE PERIODICAL

So widespread have been the criticisms of the periodical that the student will not be taken aback to hear that serious suggestions have been made to do away with it in favour of some other form of communication. The fact that most of these alternatives are open to objections even graver than those made against the periodical, and the knowledge that where any suggested replacements have been tried they have failed, lead most observers to believe that the periodical will be with us for some time yet. Nevertheless, the suggestions made are worthy of study, for individually some of them do offer remedies for individual ills. Furthermore, while in no sense supplanting the periodical, a number of the more successful innovations of recent years have very usefully supplemented it.

A large proportion of the suggested alternatives seek to make the individual paper the unit of distribution. One scheme would retain the actual papers at a central depository, circulating only abstracts to subscribers; the full papers (which could be far fuller than at present) would of course be available on request by return of post. When a plan of this kind was put as a hypothetical suggestion to the 3,021 scientists in the 1963 Advisory Council on Scientific Policy survey quoted below (page 149), a majority were in favour. In 1950 the American Society of Civil Engineers started to issue its papers as separates *instead* of in the usual bound form; some years later the Physical Society and the Chemical Society began issuing their papers as separates *in addition* to the normal bound form; both schemes failed through lack of support. More successful are the various schemes for providing preprints (not to be confused with the limited-circulation ' preprints ' described later) of articles about to appear in regular journals, but of course these do not replace the journals in any way; they merely supplement them by getting important papers into print and in circulation with the minimum of delay. The preprints are advance ' pulls ' of individual articles taken before the regular printing run; eg, *Journal of psychology* prints and publishes articles as separates on receipt, and only later are they issued as part of the quarterly volume; *Industrial and engineering chemistry* offers a similar service.[8]

In certain ways the research report functions as an alternative to the periodical article and is now of course a major source of information in many fields. It will be dealt with at length in chapter 13.

Where some positive progress has been made in the last few years is in the establishment of new types of periodicals, specifically designed to overcome one or more of the problems of conventional journals as related above. Most successful have been those attempts to reduce the normal delay in communicating new results. One simple method has been to reproduce on microfiche the authors' manuscripts (actually typescripts) just as they stand, *eg, Wildlife diseases* appears in no other form. For many years a number of journals have given previews of papers that are to appear later in full in their pages, *eg, Analytical chemistry, Biochemical journal:* there are now journals entirely devoted to such previews, *eg, Biochimica and biophysica acta previews.* In a similar way a number of journals have encouraged the rapid publication of urgent research results in brief form as ' letters to the editor ' or ' short communications ', *eg, Journal of the American Chemical Society, Nature, Science:* this has been taken a stage further by a number of titles (mostly offshoots of conventional journals) solely

8 Preprints should be distinguished from reprints, which are extra copies (usually for the author, and therefore sometimes known as author's extras) of an article, run off *after* the main printing. It is usual for scientific journals to provide perhaps 20 such reprints as a matter of course, and further copies can be ordered and paid for. Authors use these for sending to their friends, or to specialists in their field, or to those who request them (for to request such a preprint from the author is common practice in certain disciplines). Some libraries, particularly in biology, collect such reprints on a very large scale, *eg,* the Royal Botanic Gardens, Kew, has 120,000; the Royal Entomological Society has 40,000. They are of course only *reprints* of what is already in literature, but in a well-organised collection can be much more convenient of access.

given over to such rapid preliminary communications, *eg*, *Electronics letters*, *Chemical communications*, *Life sciences*, *Biochemical and biophysical research communications*. Such periodicals certainly allow speedy publication in unedited form (in the trade they are jocularly known as *Acta retracta*): *Tetrahedron letters* claims to publish within four weeks, and *Chemical physics letters* within fourteen days of acceptance. Criticism has been voiced that this facility is 'abused by publication-hungry scientists whose future promotion and grants depend on the quantity of their publications, since this is more easily assessed than quality'.

One of the most controversial alternatives to the periodical has been the attempt in various forms to revive individual scientist-to-scientist communication. Not that this has ever died out completely: so long as the function of science and technology is to push back the frontiers of our knowledge of the intricacies of nature and the application of its secrets to our environment, there will always be a small group of pioneers at these frontiers, in close personal touch with one another. But according to D J de Solla Price, 'There now exist dozens of what we call invisible colleges,[9] each consisting of the few hundred persons who make up the international body of real leaders in their subjects. They are power groups, albeit often unwittingly, and the more power they have the more they gain.' Part of this power derives from their custom of circulating within their closed group what are misleadingly called 'preprints' reporting their recent work. Their aim, to spread the good word with the minimum of delay, is entirely laudable, and of course as the most prominent workers in their various fields they do generate much of the significant new information. Unfortunately, these so-called 'preprints' have roused the ire of the editors and publishers of the journals, of librarians, and of course of those scientists outside the magic circle. Not to be confused with real preprints, which as described above (page 115) are actual printed copies of articles, taken off before the main run, these circulated papers may never in fact appear in print at all (although some of them obviously will). They are often preliminary, conjectural, tentative reports, distributed as much for comment and criticism as for information. Obviously they do not pass through the normal refereeing process, and by their nature cannot be subjected to any form of bibliographical

---

[9] So-called by analogy with the original 'Invisible College' formed by the seventeenth century pioneers who later founded the Royal Society, *eg*, John Wilkins, Robert Boyle, Sir Christopher Wren.

control. Unfortunately, they are collected, used, and sometimes cited (often in the form 'personal communication'). Herbert Coblans of ASLIB refers to this kind of 'preprint' as 'that bibliographical freak misbegotten out of war restrictions by editorial slowness, and nurtured by human vanity', and the editor of *Physical review letters* considers that 'plans to add unrefereed preprints to the mainstream of scientific communication will further increase storage of worthless information'. Many would go further, seeing in this reversion to 'the privacy of the seventeenth century' the beginning of the breakdown of the basic science communication system: in his trenchant way D J de Solla Price says: 'I think it means that the old-style function of the scientific paper is a dead duck'.

An interesting attempt was made in 1961 to formalise arrangements of this kind in the field of the biomedical sciences by establishing a number of Information Exchange Groups, with a secretariat at the US National Institutes of Health taking the responsibility for duplicating and distributing the papers to each member of the group. Known as *IEG memoranda*, these communications took a number of forms, *eg*, letters, requests for information, protests, papers already submitted to the journals and awaiting publication: a typical group (IEG no 1: Electron transfer and oxidative phosphorylation) had 725 members from 32 countries, and circulated in six years some 800 *Memoranda*, including an estimated 90 percent of all the important papers in the field. They were not available to libraries, or to scientists outside the group. Not surprisingly, this experiment soon ran foul of the established journals, understandably aggrieved at being expected to publish articles that several hundred of the top people in the field had already seen, perhaps many months before. In 1966 their editors issued an ultimatum that they would no longer accept papers previously circulated through the Information Exchange Groups. Bitter words were exchanged for several months, the groups being accused of 'an offence against scholarship' and the editors being charged with their 'unwillingness . . . to face up to the obvious facts that the journals can satisfy only a diminishing part of the needs of the scientific disciplines in respect to communication of information'. It was decided to terminate all the groups by February 1967.[10]

Similar in some ways to *IEG Memoranda* insofar as they are privately circulated are research newsletters. Like the *Memoranda*

10 David Green 'Death of an experiment' *International science and technology* 65 May 1967 82-8.

they are found mainly in the biological sciences, *eg*, *Human chromo-some newsletter*, *Laboratory primate newsletter*, *Microbial genetics bulletin*, *Mouse newsletter*, but they differ in their general availability: hardly any of them have a restricted circulation, the National Lending Library has an extensive collection, and though not published, a number are included in the *World list of scientific periodicals*. Commonly they contain social news, addresses, technical notes, brief research reports, and particularly valuable bibliographies. As basically *news* media they do not compete with the primary journals and have not encountered the same criticisms as the IEG.[11]

LISTS OF PERIODICALS

Mention of the *World list* is a reminder of the amount of effort expended, largely by librarians, on bibliographies of periodicals. Bibliographical control of the contents of periodicals through indexes and abstracts will be discussed in the next chapter: here attention is concentrated on the first stage of control—the identification and location of periodicals in science and technology.

Whilst relying heavily on such general bibliographies (with which the student will already be familiar) as *Union list of serials*, *Willing's Press guide*, the British Museum *General catalogue of printed books: periodical publications*, and especially those lists like *New serial titles— classed subject arrangement* and Mary Toase *Guide to current British periodicals* (Library Association, 1962)—both arranged by the Dewey Decimal Classification—which can be used as subject lists, the librarian can also turn to a range of bibliographies devoted to scientific and technological periodicals.

The most useful lists of current titles are the holdings lists of the major scientific and technological libraries, *eg*, National Lending Library for Science and Technology *Current serials received* (HMSO, 1967), Patent Office *Periodical publications in the Patent Office Library: list of current titles* (third edition 1965), Science Museum *Current periodicals in the Science Museum Library* (ninth edition, 1965). As such, of course, they also locate copies (in the case of the NLL, lending copies) of the journals listed. The standard retrospective list, with almost nine thousand titles, is H C Bolton *Catalogue of scientific and technical periodicals, 1665-1895* (Washington, Smithsonian Institution, second edition 1897). This is usefully supplemented by S H Scudder

[11] H V Wyatt ' Research newsletters in the biological sciences ' *Journal of documentation* 23 1967 321-7.

Catalogue of scientific serials . . . 1633-1876 (Cambridge, Mass, Harvard University, 1879), which includes the transactions of learned societies (omitted by Bolton). It excludes technology, but interestingly arranges its 4,400 titles by country and town of origin.

All these lists are comprehensive in intention, but there are a large number of deliberately selective lists using various criteria of selection, eg, *Ulrich's International periodicals directory: volume I, Scientific, technical, medical* aims to list the most important titles; in Library of Congress *Scientific and technical serial publications of the Soviet Union, 1945-1960* (Washington, 1963) and *Directory of Indian scientific periodicals* (New Delhi, INSDOC, second edition 1968) the titles are chosen on a national basis; Midwest Inter-library Center *Rarely held scientific serials* (Chicago, 1963) and Royal Society *A list of British scientific publications reporting original work or critical reviews* (1950) illustrate two other bases of selection.

Commonly encountered are the many union lists of library holdings: these may be local, *eg*, Sheffield Interchange Organisation *Union list of periodicals* (1967) with 4,500 titles; or national, *eg*, *Union list of scientific serials in Canadian libraries* (Ottawa, National Research Council of Canada, second edition 1967) with 38,000 titles.[12] These two examples remind us that the routine bibliographical task of compiling a union list is well-suited for mechanisation: the Sheffield list was compiled with the aid of punched cards and the Canadian work was computer-produced. A by-product of bibliographical mechanisation have been the schemes to devise a distinctive abbreviation (or more precisely, code) for each periodical title using as few letters as possible to facilitate computer operation. Best known are the American Society for Testing and Materials four- and now five-letter codes, *eg*, SORR-A (*Sound recording and reproduction*), EXCE-A (*Excavating engineer*). Although little more than the titles are included, the lists of these codes can serve as sort of bibliography: ASTM *Coden for periodical titles* (Philadelphia, second edition 1966) in two volumes with *Supplement* (1968), for instance, includes 55,000 titles. The results of modern technology can also be seen in the London School of Hygiene and Tropical Medicine *Serials catalogue* (Boston, Hall,

12 The *World list of scientific periodicals* (Butterworths, 1963-5) does indeed live up to its name in including periodicals from all over the world, but as a *union list* it is a national not a world list as it only indicates holdings in British libraries. However, like all union lists this work has a bibliographical value over and above its use as a location tool.

1965), a reproduction by the photolitho-offset method described earlier (page 91) of the 6,000 periodical title cards in the library catalogue.

As this example indicates, there are separate lists of periodicals in the special subject fields within science and technology. A useful guide to these and similar works (a bibliography of bibliographies, in fact) is M J Fowler *Guides to scientific periodicals: an annotated bibliography* (Library Association, [1966]). The student will find that such lists fall into the same categories as those of wider subject scope described above, namely, lists of a library's current holdings, *eg,* Geological Society of London *List of periodicals currently taken by the library* (1962) ; or retrospective lists, *eg,* Library of Congress *Aeronautical and space serial publications: a world list* (Washington, 1962); or selective lists, *eg,* Association of College and Research Libraries *A recommended list of basic periodicals in engineering and the engineering sciences* (Chicago, 1953); or union lists, *eg,* ASLIB Textile Group *Union list of holdings of textile periodicals* (third edition 1962). Some of the most useful current lists are those issued by abstracting services to let their users know which periodicals they scan, as for instance the list issued in 1968 by Biological Abstracts. One of the largest and best known of these used to be the Chemical Abstracts *List of periodicals abstracted,* but since 1969 this has been replaced by *Access,* an expansion in two volumes of 1,500 pages, produced by computer. It includes all chemical titles (not merely those current) together with locations (including back files) in libraries in the US, Canada, and in a limited number of other libraries.

Not all lists of periodicals are published separately: a number of valuable examples have appeared in periodicals, *eg,* M M Rocq and others 'Petroleum periodicals' *Special libraries* 36 1945 376-91. Similarly, the lists of periodicals abstracted by many (probably most) services appear only in the abstracting journal itself, *eg,* in the January issue each year of *International abstracts of biological sciences,* or, as in the case of *Engineering index* in the annual volume as 'List of periodical and serial publications reviewed '.

John Maddox ' Journals and the literature explosion ' *Nature* 221 1969 128-30.

A Neelameghan ' Periodicals and science communication ' *Annals of library science* 9 1962 182-8.

B C Vickery ' Bradford's law of scattering ' *Journal of documentation* 4 1948 198-203.

R H Phelps and J P Herlin 'Alternatives to the scientific periodical ' *UNESCO bulletin for libraries* 14 1960 61-75.

IO

INDEXING AND ABSTRACTING SERVICES

Once their short period of currency is over and they have become back files or bound volumes, most journals would remain closed books were it not for those keys to their contents, the indexes. Most of the more responsible periodicals with material of lasting value in their pages do attempt to publish an index to their contents, commonly once a year (or once per volume, if this is not yearly), although there is a great variety of forms to be observed. Probably the most convenient is for the index to be included with the final issue of the year (or volume), printed either as an integral part or separately. Some journals need longer to compile their indexes and have to adopt the practice of including them with a subsequent issue. More frustrating is the index which is not automatically supplied to subscribers but has to be ordered specially, and sometimes paid for as an extra. Each periodical is a law unto itself in this matter, although a number of the bibliographies of periodicals do indicate the practice of individual journals where known, *eg, Ulrich's International periodicals directory,* and there is one bibliographical tool designed specifically for this problem: the *TPI list: a check list on the title pages and indexes of . . . periodicals* (Hafner, 1961).

Even more valuable are the cumulated indexes, covering not just one but ten, fifty, or even more years of a periodical, *eg, Analyst,* volumes 1-20 (1877-96) and every ten years since; *Annals of applied biology,* volumes 1-50 (1914-62); *Engineer,* 1856-1959; Institution of Mechanical Engineers *Brief subject and author index of papers . . . 1837-1962.* Keys of this kind can turn a run of a journal into a valuable reference source.

There are many journals, however, which do not provide any index at all, and many more which provide no more than an author index. When queried, in some cases their editors will maintain that the effort (and the expense) is not justified; in others the point will be made that their particular periodical has a current value only. In some ways more disturbing, because more misleading, are inadequate indexes. Far too common is the index which through half-heartedness, technical incompetence or misguided policy, serves as a guide to only part of a journal's contents. The danger is of course that the searcher will assume that if what he is looking for is not in the index it is not in the journal. As Staveley has said: ' Some indexing services could certainly be dispensed with, and the work of others would be lightened, if journals were themselves indexed fully and systematically. It is surprising that so many editors continue to neglect this elementary duty.'

INDEXING SERVICES

For many years, among the most important bibliographical tools for controlling the periodical literature of science and technology have been those indexes which have analysed the contents not just of one but of a wide range of titles. The student can gain a valuable insight into the operation of typical current services of this kind by an examination of two parallel but contrasting monthlies in the same subject field, *Biological and agricultural index* and *Bibliography of agriculture*. The former, founded in 1916 and known for its first 49 years as *Agricultural index*, is an alphabetical subject index to about 150 periodicals. Published by the famous H W Wilson Company, the world's largest publisher of indexes for libraries, it is a characteristic example, with the monthly issues cumulating annually (and in the case of the earlier *Agricultural index* every three years as well). Following regular Wilson practice the titles indexed (all in English, and mainly from the US) are chosen by the subscribers themselves in a poll conducted at intervals by the publishers, and price is determined according to ' the service basis method of charge, based on the principle that each subscriber should pay in proportion to the amount of service used '. *Bibliography of agriculture*, on the other hand, is an official US Government publication of the National Agricultural Library. Starting in 1942, it has attempted to index not merely periodical articles but all literature, both domestic and foreign, received by the library. Arrangement is in classified *ad hoc* order, with author, corporate author and subject indexes in each issue. The December number each year is devoted solely to a cumulation of the monthly indexes. An average of close on seven thousand citations per issue makes it about twice the size of *Biological and agricultural index*. The price to subscribers is kept very low.

Until a few years ago, it would have been true to say that these two services, despite their many differences, were straightforward conventional indexes. The last decade, however, has seen a revolution in the application of machines to indexing. In *Bibliography of agriculture* the changes recently introduced and still to come are a demonstration of this revolution in practice. In 1962 an internal Task Force was set up in the National Agricultural Library to study how mechanisation could meet the anticipated growth of the *Bibliography,* and as a result year by year a series of steps has been taken towards an automated system. Starting with the monthly author index, and the annual cumulation, and progressing to the subject indexes, computer sorting and print-out based on input from optical scanning of typewritten text is now in operation.

Of course the best known and most thoroughgoing mechanisation of an indexing service is to be seen at the corresponding library in the medical field, the (US) National Library of Medicine. The MEDLARS story is too well known to be told again here[1], but the heart of the system was the complete computerisation in 1964 of *Index medicus,* founded in 1879 and the world's largest index in any subject field. With (at that time) some 13,000 citations per month, it was obviously ripe for mechanisation. At a cost of about $3 million and 30 man-years of programming labour the computer took indexed citations in the form of paper tape, and within five days (compared with over three weeks by manual methods) produced via the phototypesetter the 600-odd printed pages of the *Index.* Since then of course, the system has been refined and extended, and an upgraded MEDLARS II is in prospect.

The student will be aware that mechanisation in this context (and up till now in the context of any other indexing service apart from the primarily experimental) relates to the clerical process of indexing, *eg,* arranging, sorting, cumulating, printing-out. Progress in applying machines to the intellectual tasks involved in indexing is very slow.[2] Each of the papers indexed for MEDLARS takes between ten and fifteen minutes of the time of a graduate literature analyst.

Characteristic of indexing services for science and technology (but not of abstracting services) is the attempt to cover the whole field. According to Bottle, 'The dream of a comprehensive multi-disciplinary

---

[1] One of the best of many accounts is Leonard Karel and others ' Computerized bibliographic services for biomedicine ' *Science* 148 1965 766-72.

[2] C D Batty ' The automatic generation of index languages ' *Journal of documentation* 25 1969 142-51.

bibliographic index is probably almost as old as librarianship itself, yet the ever increasing flood of literature of all types and subjects must doom any such project to failure '. One cannot deny that the path is strewn with the remains of failed indexes of this kind (*eg, Index of technical articles, Cleaver-Hume technical article index*), but we do have the example of *Applied science and technology index,* which under its earlier title *Industrial arts index* has been covering a very wide range of disciplines since 1913. Apart from its highly professional approach (it is one of the oldest of the H W Wilson subject indexes), it probably owes its success to the care taken to respond to the needs of its users in choosing the 227 periodicals to be indexed. Obviously, these make up only a fraction of the total journals, and all are in English (mostly published in the US), but librarians can testify to its value in libraries of all kinds.

Although established only since 1962 the corresponding *British technology index* has many features of interest to the student. Like a number of other indexes it has been fully computerised in stages over the last few years.[3] Even more significantly, however, it has served as a practical focus for a second revolution in indexing to parallel the computer revolution mentioned above. This of course is the great surge of interest in subject-indexing theory over the last decade or so. Compared with earlier examples, BTI uses a very sophisticated classificatory method of indexing, with headings consisting basically of a string of subject terms, together with inversion cross-references, and synonym and relational cross-references. This approach contrasts interestingly with the *Index medicus* method, also designed in the light cast by the latest research into indexing theory: subject headings and cross-references are drawn from MeSH (Medical subject headings), a 7,000-term thesaurus published annually as Part 2 of the January *Index medicus.*

The latest of these multi-disciplinary indexes is *Pandex: current index to scientific and technical literature.* One of the new generation of indexing services inasmuch as it has been planned from the start as a computerised system, it appears in conventional form every two

[3] Once again, of course, it is only the clerical procedures that have been mechanised: ' The production of *British technology index* is *human* based, and there is no intention in the foreseeable future to try to dispense with the human intellectual effort which we believe to be necessary to achieve the standard of retrieval performance which we have set ourselves ' E J Coates ' The computerisation of the *British technology index* ' Bernard Houghton *Computer based information retrieval systems* (Bingley, 1968) 45-63.

weeks and claims to cover 1,900 journals, 6,000 books, 5,000 patents, and 35,000 research reports each year. Citations are automatically arranged under anything from six to twenty subject headings, which are in fact keywords selected from the titles of the articles and edited manually. Each issue has an author index. Cumulations on microfilm and microfiche are published quarterly and annually. Total computerisation of this kind allows a number of other services to be offered also, *eg,* 'A subscriber to the *Pandex* Weekly Magnetic Tape Service receives the total weekly input of bibliographical data for use on his own computer. *Pandex* provides complete programs for printout, retrospective search, and SDI (Selective Dissemination of Information), as part of his subscription . . . For individuals or groups interested in receiving only that information which is relevant to their particular interests or to a particular subject, *Pandex* is capable of providing individualized search programs. On a weekly, bi-weekly or monthly basis, *Pandex* can perform a comprehensive search of the current bibliographical data and identify all entries of interest to the subscriber. The information is immediately made available to the subscriber in whatever form he finds most convenient'. Services of this kind have not been available long enough or tested widely enough to permit positive estimates of their worth (compared with conventional services they are very expensive), but they certainly point the way forward. For many years yet it is likely they will be offered alongside the printed versions, but some prophets have foretold that the computer tape will eventually replace the printed page.

To the aid of the human indexer the computer brings nothing more than greatly increased manipulative power. It is this power which allows him to dispense with much of his clerical help and still to produce his index in a tenth of the time. But it also permits him to consider methods and processes that he would otherwise dismiss as far too tedious and time-consuming for even the most junior human hand. This is the explanation for the revival of interest in rotated indexes and citation indexes. Both have been known for many years, but are exceedingly laborious to compile. In a rotated index (used as long ago as 1864 by Crestadoro in the printed catalogue of the Manchester Public Library) each of the keywords in a title serves as the subject heading in its turn, the title being printed as many times as there are keywords: in its commonest computerised form, the printout centres these keywords down the page with the rest of the title (still in the same order) ' wrapped round '. Its great advantage is that its compilation needs no 'intellectual', *ie*, indexing, effort at all, and some of its recent success is probably due to shrewd publicity combined with a brilliant choice of acronym in KWIC (keyword in context). Much criticism is heard about the actual physical difficulty of using such indexes, and more basic doubts are frequently expressed on the reliability of the title alone as an indicator of a paper's content, but so speedily can they be produced that they are now widespread.[4] A typical example to study is *Bioresearch index.*

A citation index is a list of cited articles, under each of which is a further list of documents where they have been cited. This system has been used in the field of law for many years in the US (*Shepard's citations*), so all that is novel is its extension to science and the use of computer processing. Again, the great advantage of this method is that its compilation is a purely clerical operation, and no subject indexing as such is done. The author of a paper serves as his own indexer: each time he provides a citation in his paper he is indexing some aspect of his own work and re-indexing the scientific literature. The success of a citation index depends of course on the extent to which items cited in the bibliographies to papers reflect the contents of those papers. In searching a special technique is necessary: before he can start, for instance, the user must have a starting point; he must know at least one article on the subject of his search. By far the best known is *Science citation index,* covering some 1,800 journals containing about 300,000 articles per year. These 'source' items generate something over three million cited references. To manipulate this data manually would be a superhuman task, but the computer produces not only the quarterly issues of *Science citation index* itself but the companion *Source index* (authors and titles of citing articles) and *Permuterm subject index* (a pre-coordinated index showing the permutations of all possible pairs of terms derived from each article).[5]

An indication of the importance attached by scientists to the periodical literature and its bibliographical control is the effort expended on retrospective indexing, giving science advantages still not shared with most other disciplines. Outstanding is the Royal Society *Cata-*

---

[4] Marguerite Fischer ' The KWIC indexing concept: a retrospective view ' *American documentation* 17 1966 57-70.

[5] John Martyn 'An examination of citation indexes ' *ASLIB proceedings* 17 1965 184-96.

M V Malin ' The *Science citation index*: a new concept in indexing ' *Library trends* 16 1967-8 374-87.

logue of scientific papers, 1800-1900 (Clay, 1867-1902; Cambridge UP 1914-25), covering in its 19 volumes over 1,500 nineteenth century periodicals. This is an author index only, however: the complementary *Subject index* (Cambridge UP, 1908-14) was abandoned incomplete after only three of the projected 17 volumes. Also abandoned was the successor to this venture, the ambitious *International catalogue of scientific literature* (1902-19), designed 'to record the titles of all original contributions since Jan 1, 1901'.

The fact that this last title included books as well as papers in journals reminds us that many indexes, current and retrospective, cover more than just periodicals. As we have noted (page 92) they combine the features of an index (*ie*, of periodical articles) and a bibliography (*ie*, of books), *eg*, the annual *Index to the publications of the Iron and Steel Institute*, the annual *Zoological record*, Bashford Dean *Bibliography of fishes* (New York, American Museum of Natural History, 1916-23), with 35,000 titles, 'all the published references to fishes', in three volumes. And occasionally to be encountered is the composite book index, *eg* L J Fogel *Composite index to marine science and technology* (San Diego, Calif, Alfo, 1968), a collation of the indexes of 30 books.

ABSTRACTING SERVICES

To fit the definition of an indexing service, all that is necessary is for the citation to provide sufficient bibliographical information about each item to enable it to be identified and traced. In practice this means that the scientific or technological information content of a reference is limited to the subject-heading chosen and the title (or to the title alone in title-based indexes such as KWIC). It is probably true that titles in science and technology are on the whole more informative and explicit than in other fields; some indexes do 'enrich' titles with words and phrases added by the indexer; and descriptors accompanying a citation can sometimes be pressed into service as a 'skeleton' summary of subject content. Nevertheless, as sources of information, indexes have obvious limitations. Abstracts, however, are archetypal secondary information sources: comprising not merely citations but also summaries of the content of publications or articles, they manifestly 'organise the primary literature in more convenient form'. As a tool for the scientist or technologist the abstracting service is double-edged: not only does it alert him (as an indexing service does) to newly-published work that the law of scattering has so dispersed that he would without its aid miss completely, but it can often obviate the actual perusal of the original journals. Faced with the steadily rising tide of primary publications, research workers not surprisingly grasp eagerly at publications which save them time. And retrospectively, as a repository in summary form of the literature in its field, an abstracting service permits retrieval of specific information. Indeed, many writers lay great stress on what might be called this ' encyclopedic ' function, *eg,* ' The index of an appropriate abstracting journal should be used whenever information is being sought on any subject, and it is likely to lead to information of more value than would be obtained from many other books of reference ' (ASLIB *Handbook*).

A distinction is commonly drawn between an *indicative* abstract (' a brief abstract written with the intention of enabling the reader to decide whether he should refer to the original publication or article '), and an *informative* abstract (which ' summarizes the principal arguments and gives the principal data in the original publication or article '). In practice there are many abstracts published which do not fall clearly into one or other category, and many services (a quarter of 130 surveyed in 1962) publish both, *eg, Horticultural abstracts, Building science abstracts.* In any case it is probably unwise to rely solely even on very long informative abstracts: any serious worker would also consult the original. Most abstracts, whether indicative or informative, are merely descriptive, with no attempt at evaluation. Indeed, abstractors are commonly instructed to avoid such an assessment, *eg,* 'An abstract should be impersonal . . . and should not take critical form . . . the only acceptable criticism is that of giving very little space to the abstract, or of ignoring the article entirely if it is definitely lacking in value '. Of particular interest therefore are those few services which do attempt evaluation, either in the text of their abstracts, *eg, ANBAR abstracts,* and a number of medical abstracting services; or by their policy of selecting only ' some of the more important and interesting recently published papers ', *eg, Chemistry and industry, Food manufacture.*

FORM OF ABSTRACTS

In contrast to indexing services, a very large number of abstracting services are not separately published but appear as a feature within a particular journal, *eg, Vacuum, Journal of the science of food and agriculture, Journal of the Institute of Brewing, Ultrasonics, Glass* technology, *Production engineer*. This, of course, is a practice far older than the pure abstracting journal, the earliest of which is thought to be *Pharmaceutisches Centralblatt* (1830). Sometimes the bulk of the periodical is given over to abstracts, *eg, Journal of applied chemistry*, and examples of abstracts are increasingly to be found in review journals, *eg, Metron : Sira measurement and control abstracts and reviews*.

But the most obvious form for an abstracting service is a journal devoted to abstracts, *eg, Dairy science abstracts, Lead abstracts, Indian science abstracts, Computer abstracts*. Some of them aim at comprehensiveness, trying to abstract every publication or article appearing in their subject field which contains valuable or original material, *eg, Fuel abstracts, British ceramic abstracts*. Others are deliberately selective, and include only those items which they think are major contributions to knowledge, or are likely to be of use to a particular class of reader, *eg, Abstracts of world medicine*. A variant of the abstract journal is the abstract annual, issued in the form of a bound volume, *eg, Gas chromatography abstracts*.

A format very common on the Continent of Europe but less frequently encountered in Britain and America is the abstracting service on cards. A trifle unusual in that it uses 6in × 4in cards rather than 5in × 3in is *Training abstracts;* even more striking is the American Society of Civil Engineers *Publications abstracts* with removable cards in journal form. A number of services provide cards as well as a conventional journal, *eg, Engineering index* (an abstracting service despite its title), *IAG—new literature on automation*. Several journals provide the wherewithal to make your own cards by printing their abstracts on one side of the paper only ready for mounting (and in some cases already gummed), *eg, Journal of animal ecology*. This format is often used for the so-called 'homotopic' abstracts (again more common in Europe): these are usually printed on a sheet inserted near the front of a journal and comprise abstracts (often by the author himself) of the actual articles in that issue, *eg, Glass technology*, where the abstracts are in French and German as well as English.

Frequency of publication obviously can vary from daily upwards, with perhaps monthly being the commonest. Cumulations (of the actual abstracts, as opposed to the indexes), are rare, but not unknown: a very well-known instance is *Engineering index*, a monthly which cumulates annually.

There are probably about two thousand abstracting and indexing services currently available in science and technology: 1,855 of these were listed in 1963 in *A guide to the world's abstracting and indexing services in science and technology* (Washington, National Federation of Science Abstracting and Indexing Services), and the National Lending Library has published *A KWIC index to the English language abstracting and indexing publications currently being received* (second edition 1967). The student could never hope to familiarise himself with more than a fraction of these, but like periodicals, for purposes of study abstracting services can usefully be grouped according to their source of origin, as follows:

a) *Learned societies, professional bodies, eg, Photographic abstracts* (Royal Photographic Society); and *Applied mechanics reviews* (American Society of Mechanical Engineers) and *Mathematical reviews* (American Mathematical Society), both of which are abstracting services despite their titles.

b) *Governmental bodies*: the state is often responsible for abstracting services issued by its various organs such as ministries, departments, commissions, boards, etc, *eg, Index aeronauticus* (Ministry of Technology), Ministry of Public Building and Works *Library bulletin, Road abstracts* (Road Research Laboratory, Ministry of Transport), *Water pollution abstracts* (Water Pollution Research Laboratory, Ministry of Technology), *Nuclear science abstracts* (US Atomic Energy Commission), *CSIRO abstracts* (Commonwealth Scientific and Industrial Research Organisation). More commonly in Britain, state support of abstracting services takes the indirect form of a grant-in-aid to bodies such as industrial research associations or specialised information centres, which often publish abstract journals as part of their function *eg, Monthly summary of automobile engineering literature* (Motor Industry Research Association), *Journal of abstracts* (British Ship Research Association), *Ergonomics abstracts* (Ergonomics Information Analysis Centre). A feature of recent years has been the number of cooperative ventures in abstracting, *eg, World textile abstracts,* a combined operation by five research associations in the textile field; *Food science and technology abstracts,* a computer-produced journal prepared jointly by the Commonwealth Agricultural Bureaux (UK), Institute of Food Technologists (US), and Institut für Dokumentationswesen (Federal Republic of Germany).

The student will have noticed that a number of these governmental abstracting services are what is called ' mission-oriented ' as opposed to ' discipline-oriented '. It is a fact that the organisation of abstracting services reflects the organisation of science itself. Large-scale public funding (*eg*, defence, space, nuclear energy) has produced a situation where much scientific and technological research is ' directed to the solution of complex problems in a society rather than to the advance of knowledge in an academic field '.[6] Abstracting services oriented to missions are characteristically interdisciplinary.

In certain countries it is found that the state itself operates specialised abstracting organisations, *eg*, Centre National de la Recherche Scientifique (Paris) compiles *Bulletin signalétique;* VINITI (Moscow) compiles *Referativnyi zhurnal.*

c) *Independent research institutes*: these are not common, and are usually confined to topics of current interest to the institute, *eg*, *Selected Rand abstracts* (Rand Corporation).

d) *Commercial publishers*: so laborious and thankless a task is abstract compilation that most publishing houses steer clear, but there are exceptions, *eg*, *Deep sea research and oceanographic abstracts.*

e) *Industrial and commercial bodies*: abstracts issued by individual firms are usually produced by their library or information departments for internal consumption only, but many do circulate more widely, *eg*, *Rolls Royce bulletin, Titanium abstract bulletin* (ICI Ltd, Metals Division). Trade development associations are active in this field also, *eg*, *Copper abstracts* (Copper Development Association), *Zinc abstracts* (Zinc Development Association).

f) *Specialised abstracting services*: a number of the larger abstracting services are run as independent organisations, sometimes incorporated. Theoretically self-sufficient they may on occasion receive financial support from the state, *eg*, *Biological abstracts.*

g) *Libraries and information services*: as the student will have observed in the examples noted, the libraries of learned societies, government departments, research associations, industrial firms, etc, frequently take or share responsibility for the abstracting service issued by their parent bodies.

It would be inappropriate to the purpose of this textbook to describe in detail any particular abstracting services, but it is essential

[6] Scott Adams and D B Baker ' Mission and discipline orientation in scientific abstracting and indexing services ' *Library trends* 16 1967-8 307-22.

for the student wishing to appreciate their role in scientific and technological communication to devote some time to the actual physical
examination of major services in their basic printed form, eg, *Chemical
abstracts, Biological abstracts, Science abstracts*. From this he should
then move on to a broader study of the recent radical advances made
or proposed by these services. To some extent these developments have
been forced by the increasing inadequacy of manual methods in the
face of the ever-growing flood of publications, and typically take the
form of a total computerised system, not only producing the basic
printed abstracting journal but offering a range of other services also,
eg, CAS (Chemical Abstracts Service),[7] BIOSIS (BioSciences Information Service),[8] and INSPEC (Information Service for Physics, Electronics and Control).[9]

COVERAGE OF THE LITERATURE BY ABSTRACTING SERVICES
Estimates prepared at the National Lending Library indicate that the
26,000 scientific and technological periodicals current in 1965 contained approximately 850,000 authored articles. Other estimates have
ranged as high as four times that figure, but none have approached
7·5 million, which is a reliable estimate of the number of references
produced by the indexing and abstracting journals. Clearly duplication
must be widespread. In a classic series of statistical tests carried out
for ASLIB it was shown that 47 percent of a sample of 3,420 references
were abstracted more than once, and in some subjects 22 percent were
covered four times.[10] It is sometimes advanced as justification for such
overlapping that each abstract is made from a particular viewpoint,
and a summary of an article prepared for a chemist, for instance, will
not serve the needs of a biologist. There is some truth in this, and
the Commonwealth Agricultural Bureaux, for example, responsible
for over a dozen abstracting services, often prepare three or four
abstracts of the same paper deliberately, each with a different subject-
slant. But this is not why there is so much duplication: the ASLIB

[7] American Chemical Society *CAS today* (Columbus, Ohio, 1967).

[8] Louise Schultz ' New developments in biological abstracting and indexing '
Library trends 16 1967-8 337-52.

[9] T M Aitchison 'A computer-based information service in physics, electro-
technology and control ' Institute of Information Scientists *Proceedings of
the third conference . . . Sheffield . . . 1968* (1969) 88-99

[10] John Martyn and Margaret Slater ' Tests on abstract journals ' *Journal of
documentation* 20 1964 212-35 and 23 1967 45-70.

tests showed very little evidence of genuine slanting in duplicate abstracts, and an earlier US investigation of papers covered twice or more in nine abstracting services showed that most of the abstracts were in fact written by the authors of the papers themselves, and in a very large number of cases they had appeared with the original papers. The widespread acceptance of such author abstracts by respected abstracting services does indicate that subject-slanting is not always necessary, and that it is possible to produce a useful 'neutral' abstract.

A far more disturbing problem of coverage is omission. The ASLIB tests showed that 21 percent of the 3,420 references were not covered by the abstracting services, supporting an earlier US survey that had concluded: ' . . . other techniques must be combined with the judicious use of abstract journals and indexes for the greatest possible efficiency '. In 1964 an examination of coverage of the electrical engineering period-ical literature found similarly that ' The published abstracts journals do not abstract material completely enough . . . to make our own [*ie*, internal] indexes unnecessary '. In 1962 a writer on the ' Information crisis in biology ' painted an even gloomier picture: ' The biological literature which is abstracted and indexed is less than one quarter of that published '. It is probable that the new computerised systems in both these fields have improved coverage, but there is still no solution yet to one interesting problem discovered by the ASLIB team: ' . . . we have been unable to identify any general reason why material is not covered by abstracts services. There is certainly no evidence that the material not abstracted is irrelevant or of lower quality.'

The student should distinguish when examining abstract journals between nominal and actual coverage: a number of services cover their listed journals only selectively, not to say haphazardly. He should also remember that some services exclude some forms of literature, *eg*, books, patents, theses, research reports, conference papers, etc. And of course Bradford's Law of Scattering (page 113) is highly relevant here: one could scarcely wish for a more precisely defined subject than turtles, yet we are told that the literature of turtles appears in 600 journals.

INDEXING OF ABSTRACTS

Most indexing services are self-indexing by reason of their usual arrangement in alphabetical order of subject. Most abstracting services,[11] on the other hand, have adopted another arrangement, commonly a fairly broad *ad hoc* classification, *eg, Weed abstracts*. Some follow a recognised general scheme like the Universal Decimal Classification (which is very popular), *eg, Apicultural abstracts;* some use a special faceted classification, *eg, Occupational safety and health abstracts*. A number are found with their entries arranged alphabetically by title of source periodical, *eg, ANBAR abstracts*. Obviously, if such services are to fulfil their role as retrospective retrieval systems they must be provided with indexes to permit specific subject access. The user is normally content to scan each current issue as it appears in order to keep himself alerted, but if he is obliged to search back numbers for specific information he demands a subject index.

This is highly relevant to coverage: what is not indexed (even though abstracted) will not be retrieved in a retrospective search. This is well-understood by the major services: *Nuclear science abstracts* warns its staff that 'A collection of abstracts is only as good as its indexes '. Close on half of *Chemical abstracts*' full-time professional manpower and a quarter of the operating budget are devoted to the indexing effort.

The more frequently an index is produced, the fewer unindexed issues will require page-by-page searching in the course of a typical search. Annual indexes are probably the commonest still, but mechanisation has allowed several services the luxury of more frequent appearances, and it is now possible for very little extra effort to have subject indexes in every issue (although most of these are usually no more than keyword indexes). Similarly, the manipulative power of the computer permits author indexes, patent number indexes, organisation indexes, etc, from the same data base. The other side of this particular coin, however, is the multiple sequences facing the searcher pursuing an exhaustive search. Cumulated indexes are obviously of value here, *eg, Gas chromatography abstracts*, 1958-63, *Nuclear science abstracts* every five years, and the immense *Chemical abstracts* indexes (28 volumes and 40,000 pages to cover 1962-6!)

The student will know that the efficiency of subject indexes is one of the topics most actively investigated at the present time. He will know that no index is 100 percent perfect, and in the words of E J Crane, for many years Director and Editor of *Chemical abstracts*, ' Even in the use of the best subject indexes the user must meet the

[11] The outstanding exception is once again *Engineering index,* arranged alphabetically by subject.

indexer part way for good results '. And there will still remain some references undisclosed. The ASLIB tests reported that a searcher would be ' unlikely to find more than threequarters [of abstracts on the subject of his search] *via* the subject indexes, and he is unlikely to be able to find more than half without the exercise of considerable ingenuity or a good knowledge of the subject '.

Tracing an appropriate abstract is usually only the first stage in a bibliographical quest, particularly if it turns out to be merely indicative, and access to the original article is commonly required. This, of course, is the task of the library, but some abstracting services do offer help in a variety of ways, *eg*, lending the original from the library (*Zinc abstracts*); offering a photocopy (*Biological abstracts*); offering issues of the original journal for sale (*ANBAR abstracts*); ensuring that copies are deposited in certain named libraries (*Nuclear science abstracts*); or simply indicating library locations (*Chemical abstracts*).

CO-OPERATION IN ABSTRACTING

The dilemma of the abstracting services is summarised with crystal clarity by the ASLIB investigators: ' The search product is the available portion, of the indexed portion, of the abstracted portion of the total relevant literature '. Part of the solution lies in co-ordinating current effort: as Wilfred Ashworth says, ' . . . there is little doubt that enough energy is already being used which would, if properly applied, give complete coverage of all literature '. Centralisation would be one method of eliminating overlap and enabling gaps to be more easily seen: such is the system operated since 1952 at the national level by VINITI in the USSR. By contrast, the USA relies on voluntary co-operation co-ordinated by the National Federation of Science Abstracting and Indexing Services, founded in 1958, and at the international level since 1952 there has been the Abstracting Board of the International Congress of Scientific Unions. An extended account of the efforts made towards better co-ordination of services would be inappropriate here, but the student can observe some positive results in the literature, *eg*, *Metal abstracts*, formed in 1969 by a merger of *Metallurgical abstracts* of the (UK) Institute of Metals and *Review of metal literature* of the American Society for Metals. The computer casts its long shadow here also, making possible and even stimulating several examples of national co-ordination and international co-operation. There are now a number of international agreements under which individual countries abstract their national literature to provide input to international systems in return for the use of the data base, *eg,* International Road Research Documentation,[12] International Nuclear Information System.[13]

'CURRENT AWARENESS' SERVICES

What the computer has not yet been able to do is to write an abstract. This of course is an intellectual process, and what John Martyn wrote in 1967 still remains true: 'It is not yet possible to produce an abstract which will summarise a document, in sentences not found in the document, by any other than human means'. It is also a slow process, responsible for much of the delay between the publication of an article and its appearance in an abstracting journal. Delay hampers the repository role of an abstracting service hardly at all, but it can cripple its alerting function. It is a telling indication of the pace of scientific and technological discovery and of the growth of the literature that over the last ten years a number of services have been started to bridge this gap between the publication of an article and its abstract, *eg, Current chemical papers,* the first (1954) publication designed specifically for 'current awareness'. The fact that a number of these have been produced by the abstracting services themselves, *eg, Current papers in electrical and electronics engineering,* 'congruent in its coverage of the literature with its corresponding abstracts journal', demonstrates the truth of Bottle's statement that 'Production delays and the time required to produce subject indexes for abstracts have almost eliminated their one-time function as a news-giving service'. The 1964 survey mentioned above (page 134) of electrical engineering abstracts discovered that six months after publication only 17 percent of a particular sample of papers had been abstracted, and only 66 percent after twelve months. As alerting services the abstracts are clearly failing.

These newer services are able to appear more rapidly because most of them are merely title *announcement* lists, arranged in broad subject classes, but not indexed, *eg, Current papers in physics.* A number are indeed title *indexes,* but depend for their promptness on mechanisation, *eg, Chemical titles,* a KWIC index produced by *Chemical abstracts.* Commonly, they claim to include papers within a month of receipt,

12 P E Mongar 'International co-operation in abstracting services for road engineering' *Information scientist* 3 July 1969 15-62.

13 J E Woolston 'The International Nuclear Information System (INIS)' *UNESCO bulletin for libraries* 23 1969 135-8, 147.

5*

but they usually concentrate on a limited number of core journals: *Chemical titles,* for instance, covers 650 out of the 12,000 monitored by the parent service.

An even simpler form of title announcement service is the contents list, requiring a minimum of preparation. This is simply a transcription of the contents of current issues of the journals within a particular subject, and is a development of the common habit of learned journals of printing the title pages of their contemporaries. Modern technology has given us in recent years a revival of this form of alerting service, using actual reproductions of contents pages printed by photolitho-offset and with computerised indexes, *eg, Current contents: chemical sciences, Contents of contemporary mathematical journals, Current contents in marine sciences.*

Such current awareness services, rightly used, have no permanent value, for the papers in due course are covered by the abstracting services. But for current awareness, there is now some evidence to show that abstracts do not offer significant advantages over such simple title lists. What we are observing in the literature is a common enough transmutation in life—differentiation and specialisation.

COMPUTERISATION OF ABSTRACTING SERVICES

As the student will have noticed throughout this chapter, discussion of present-day services is impossible without frequent references to mechanisation. At its simplest, the computer is seen as the *only* method of coping with the continued expansion of knowledge: there are, for instance, over four hundred new chemical compounds reported in the literature every working day. But many see in the application of machines to an abstracting service an opportunity to effect a complete reconstruction. Some have grasped the chance and transformed their services (often with substantial aid from public funds), producing not merely a mechanised abstracting service, but adding a new dimension: 'In a computer-based system, information selected in a single intellectual analysis of the source documents, an analysis combining both abstracting and indexing, is put into a unified machine-manipulative store through a single keyboarding. Then from the unified bank of information, material appropriate for special-subject alerting and retrieval publication can be drawn, largely by computer programs'. In addition to the printed abstracts such systems offer, for instance, *via* the computer:

a) Current awareness services, *eg, Chemical titles,* the first (1961) computer-produced journal *(Chemical abstracts).*

b) Regular bibliographies on specific subjects, *eg, Artificial kidney bibliography* (quarterly from MEDLARS).

c) Specific bibliographies resulting from file searches, *eg,* ' Rubella or German measles ' *(Biological abstracts),* ' Battered child syndrome ' (MEDLARS).

d) Individual services such as *ad hoc* custom searches or an SDI service.

Each of these is produced with a minimum of extra effort simply by ' exploitation of the machine record '. Produced with even less effort, almost as by-products of the regular service, are derivative publications such as *Abstracts of mycology* (from Biological Abstracts). We have seen too how extra or more frequent (and indeed more detailed) indexes are also quite feasible, *eg,* the CROSS index (computer rearrangement of subject specialities) with *Biological abstracts.*

The most striking advance in service, however, is the availability of the data in machine-readable form (usually magnetic tape) which subscribers can purchase instead of or in addition to the printed version and manipulate at will on their own computers. In some cases the whole of the data base *(ie,* citations, abstracts, indexes, etc) is available in this way, *eg,* COMPENDEX, which is *Engineering index monthly* in magnetic tape form. Other services offer selected packages, either on a subject basis, *eg,* POST, a *Chemical abstracts* current awareness service covering polymer science and technology; or, like *Chemical titles,* on the basis of significance. One stage beyond these are those services available only in magnetic tape form, *eg, CA condensates,* which includes all the citations and keyword index entries from *Chemical abstracts,* but not the text of the abstracts themselves : and *BA previews,* a similar service from *Biological abstracts/Bioresearch index,* but which appears one month ahead of the printed version.

A 1966 investigation of the total cost of abstracting and indexing services in the US produced an estimate of $50 million a year. It is surprising that user studies reveal a surprisingly low usage, particularly among technologists. 53 percent of mechanical engineers do not see any abstracting or indexing journal regularly. In the electrical and electronics industries one survey showed that only 38 percent of technologists are aware of abstracts in their special field, and only 31 percent claim to make use of them. An ASLIB survey of technical library use found abstracts the least productive in yielding source material, with 7 percent compared with 40 percent for periodicals and 19 percent for textbooks. Investigations of items requested from the NLL showed that (with the exception of *Chemical abstracts*) ' abstracting journals when regarded as individual sources may be relatively insignificant ': out of 9,182 references only 43 percent had come from abstracting journals. It is true that academic users and pure scientists make more use of such services, but it is still surprisingly light.

We have seen earlier how abstracts are abdicating their alerting functions to more rapidly published media. As for their retrospective retrieval function, John Martyn says, ' Ultimately, it may be wondered whether, when these large-scale search services are available to a majority of potential users via remote console access to a central computer store, there will any longer be a need for the old-fashioned printed version of an abstracts journal '.

FURTHER READING

' Periodicals: indexes and abstracts ' Ronald Staveley *Introduction to subject study* (Deutsch, 1967) 194-215.

' Published indexing and abstracting services ' Jack Burkett *Trends in special librarianship* (Bingley, 1968) 35-72.

F A Tate and J L Wood ' Libraries and abstracting and indexing services—a study in interdependency' *Library trends* 16 1967-8 353-73.

R Satyanarayana and A S Raizada ' On current awareness services ' *Annals of library science* 14 1967 152-60.

II

REVIEWS OF PROGRESS

The elaborate and irreplaceable apparatus of indexing and abstracting services described in the previous chapter can unfortunately do nothing to stem the swelling tide of primary publication. Invaluable as these keys to information undoubtedly are, it cannot be denied that they *add* to the total amount of literature, and the scientist or technologist often finds himself in sympathy with James Thurber's friend who complained in some distress: ' So much has already been written about everything that one can't find out anything about it '.

It has been clear for some years that even the scanning of indexes and abstracts is proving too much for some workers, and there have been urgent pleas for more easily digestible forms of secondary publication. In response we have seen a remarkable revival of the review, a literature form far older than the abstract, but which has lain in its shadow for a hundred years or more. Not to be confused with the book review (of the kind found in the *Times literary supplement*), it takes the form of a critical summary by a specialist of developments in a particular field of endeavour over a given period.

Such reviews of progress are now seen very definitely to be of great importance; by some they are regarded as offering a possible pathway out of the literature jungle. H V Wyatt[1] for instance considers that ' The future of biological literature lies not in classification by words but in distillation by review'. Lord Todd, Cambridge chemist and Nobel prizeman, has stated: '. . . it will, I believe, be necessary to develop review journals intensively if we are to have a really effective information system '. And of course the editors of review series are ardent apologists: the preface to the first (1962) *Advances in nuclear science and technology* refers to the ' bewildering information problem to both the expert working along its narrow crevices and the dilettantes hoping

---

[1] R T Bottle and H V Wyatt *The use of biological literature* (Butterworths, 1966) 260.

to keep abreast of the ever expanding frontiers. Clearly what is needed by both groups are well-organised review articles.' The introduction to the first (1965) *Advances in chromatography* says: ' It is clear that the individual worker, if he is to preserve even a moderate knowledge of the entire field, must rely more upon responsible surveys than on the attempt to read the avalanche of original research papers '. And neatly drawing on his own subject—a method of chemical analysis— by way of illustration, the writer goes on to explain the particular literature difficulty that the scientist has found better solved by the review than the abstracting service: ' The problem briefly stated, is one of information sorting; we wish the uses of chromatography were so universal that it could separate information—the hard core advances from the overwhelming mass of supporting evidence and data that, although necessary in research articles, quickly swamps the digestive process '. The preface to the first (1960) *Advances in computers* describes how the review is ' intended to occupy a position inter- mediate between a technical journal and a collection of handbooks or monographs. It is customary for a new scientific or technical result to appear first in a journal, in a form which makes it accessible to specialists only. Years later it may be combined with many other related results into a comprehensive treatise or monograph. There appears to be a need for bridging the gap between these modes of publication, by surveying recent progress in a field at intervals of a few years and presenting it in a form suitable for a wider audience.' That such reviews are seen as supplementing rather than supplanting the abstract journals, however, is well demonstrated by the USSR All-Union Institute of Scientific and Technical Information (VINITI), the world's largest abstracting organisation, which publishes about a hundred different series of annual reviews, based on the abstracts in *Referativnyi zhurnal.*

As is so often the case with other types of information source, there exists an area here of considerable confusion over terminology. As we have noted, the word ' review ' can also mean a book review; the dictionary definition also allows it to cover a periodical or newspaper in general, *eg, Westminster review, Quarterly review*. Although the term does appear in the titles of many of the publications we are examining in this chapter, *eg, International review of cytology, Annual review of physical chemistry, Review of coal tar technology, World review of nutrition and dietetics*, the vast majority of such works call themselves something else. By far the most popular titles are *Advances in* . . . and *Progress in* . . ., but other examples are *Reports on progress in physics*, *Surveys of progress in chemistry*, *Recent progress in hormone research*, *Survey of biological progress*, *Record of chemical progress*, *Minerals yearbook*, *Medical annual*. There are also a number of titles which give no indication that they are review series, *eg*, *Physics and chemistry of the earth*, *Vitamins and hormones*, *Oceanography and marine biology*, but even more misleading are works with review-type titles that are really something else. *Progress in fast neutron physics* (Chicago UP, 1963) actually contains the proceedings of an international conference; *Progress in nuclear energy* is the overall title for twelve series totalling some fifty volumes of a miscellaneous nature; *Progress in microscopy* is a monograph by M Francon; *Advances in chemistry* and *Mathematical surveys* are both monograph series issued respectively by the American Chemical Society and the American Mathematical Society; and *Physical review* and *Review of scientific instruments* are both primary research periodicals.

A more serious effect of this terminological pliability is that it conceals the fact that there are two distinct categories of reviews of progress. These stem from the dilemma that always faces the editor or compiler or publisher of a regular survey of a particular subject area: should he attempt an ordered and balanced sampling of the whole field, or should he emphasise those topics of the most pressing current concern? By and large, the various review series have chosen to follow positively one path or the other: unfortunately, this vagueness of vocabulary means that users cannot determine from the title alone which choice has been made. A moment's examination of the text, however, is usually sufficient to distinguish the two types.

COMPREHENSIVE REVIEWS

These are thorough, systematic, and condensed accounts of developments in a broad field over a narrow time interval (and sometimes within a particular geographical area). Long-established examples to study are *Annual reports on the progress of chemistry* (1904-), and *Annual review of biochemistry* (1931-). They are written by teams of specialists for fellow-specialists, and usually assume extensive knowledge of the subject. Firmly based on the literature, they provide extensive references, *eg*, in *Annual surveys of organometallic chemistry* for 1965 the six-page survey on aluminium has 76 references; the fifteen-page account of carpets in *Review of textile progress* for 1965-6 has 114 references. They can unfortunately degenerate at times into narrative bibliographies or, even worse, into sterile ' literature reports '. The introduction to the first volume (1965) of *Advances in chromatography* refers with some scorn to ' the glorified bibliography or reference-finding system that some reviews tend to become when the author, actively engaged in research in a very restricted area, attempts a comprehensive survey '. Indeed, one of the contributions to the 1965 *Progress in dielectrics* is a 33-page bibliography! As Mellon says, ' With the facilities now available, one familiar with index serials and abstracting journals should be able to compile his own report of progress, or at least a non-critical summary '. Of course, the better series are far more than mere compilations: not only are they evaluative, but they play a very positive role synthesising the developments newly-reported in the literature and relating them to the already accumulated knowledge in their discipline. Here we see clearly their advantage over the abstracting services, which merely report. The best of the review series go further in attempting to isolate trends and even to point the way forward.

Probably the majority of such surveys appear annually, in the form of a single bound volume, *eg*, *Reports on the progress of applied chemistry*, *Progress of rubber technology*, *Annual review of nuclear science*, although in recent years the pressure of new literature has forced the two well-known series mentioned at the beginning of this section to expand into two volumes.[2]

An alternative solution to this problem of course is to publish more frequently than once a year, *eg*, *Quarterly literature reports: polymers*. *Reports on progress in physics* can from 1969 be obtained in parts appearing in alternate months or as a bound volume twice a year.

Some surveys are published as articles in periodicals, *eg* ' Progress in heat transfer—review of current literature ', annually in *Chemical and process engineering*; 'Annual review of the literature on fats, oils, and detergents ' in *Journal of the American Oil Chemists' Society*.

Surveys of this kind have an obvious current appeal to the specialist in that their comprehensive nature enables him to fill any gaps in

[2] A questionnaire revealed that the average subscriber to *Annual reports on the progress of chemistry* claimed to read 30 percent but admitted to ignoring 30 percent of each volume: from 1968 therefore the work has been available in two separate volumes (general, physical and inorganic chemistry; and inorganic chemistry). The preface to the 1966 volume of *Annual review of biochemistry* is more waggish: ' This year our *Annual review* has shown behavior akin to a primary biological phenomenon: it has undergone binary fission '.

his knowledge of recent developments and their broad coverage can often give him a new angle on his subject. Retrospectively, they are well-suited to serve the exhaustive approach also.

TOPICAL REVIEWS

These are 'state-of-the-art' reports on selected, specific topics of active current interest. Increasingly in the last two decades these have appeared collected in volumes issued as a series, *eg, Advances in electronics, Progress in semiconductors, Reviews in engineering geology.* Examples of individual reviews in such volumes are 'Brewing—past and present' in *Viewpoints in biology* 3 (1964); 'Drugs and aggressiveness' in *Advances in pharmacology* 5 (1967); 'Jewels for industry' in *Modern materials: advances in development and applications* 6 (1968).

They are specifically designed to be intelligible to the non-specialist, and while not 'popular' in approach, are aimed at all levels of readership from the student to the director of research. One particular aim they have is interdisciplinary cross-fertilization, and their target is the worker in related fields of science and technology anxious to remain in touch with the more significant developments outside his immediate area of interest. The gravity of this problem is emphasised by Lord Todd: '. . . it is difficult to get information conveyed between one science and another, but this is usually because the practitioners of different sciences don't quite know how to express what they are looking for in comprehensible terms or in a way that will register with each other. Yet such interchange is vital to development in the borderlands between established disciplines and it is in such borderlands that the growing points of science are frequently found. I believe that only the properly written review journal, coupled with personal contacts, can solve this problem.' That such reviews of progress are achieving some success is evidenced by the reaction to the first (1960) *Advances in computers* volume, 'felt by many readers as a welcome antidote to the ever-growing specialization of technical fields'. And some have even more specific aims: *Survey of progress in chemistry* is intended to improve the transmission of new material mainly to the college chemistry teacher.

Although written by specialists, like the comprehensive surveys, these topical reviews are seen by their editors as something much more flexible. The author contributing to *Polymer reviews* is 'encouraged to speculate, to present his own opinions and theories to give a more " personal flavour " than is customary '. Likewise, the contributions to *Problems in biology* ' differ from other reviews in that their accent is on exposition. They are not intended to present a high-powered balanced account of the current factual information, instead the author has been encouraged to be selective in the choice of material and, if necessary, to present a personal view of the subject.' They are naturally (and indeed deliberately) far less ' bibliographical ' than the comprehensive type of survey, and are not limited in coverage to a particular span of time or geographical area. The editors of *Viewpoints in biology* instruct prospective contributors that their ' broadly-based reviews . . . should not only summarize the state of the subject but also indicate the direction in which progress may be expected, and stress unsolved problems. While putting a cogent, well-argued point of view the authors will, however, not necessarily be asked to give exhaustive documentations of all work in the subject.' Even more firmly the preface to the first (1963) volume of *Progress in nucleic acid research* states: ' We do not wish to sponsor an annual or fixed-date publication in which the advances of a given period of time are summarized, or a bibliographic review or literature survey. We seek rather to encourage the writing of " essays in circumscribed areas ".'

This same freedom of approach permits the author of a typical review if he so wishes to write at greater length than is usual in a conventional periodical article, without having to produce a full-blown monograph. However, if a contributor feels he must produce a work of such substance, there is evidence that the review format is hospitable enough to accommodate him: the fifth (1966) *Advances in marine biology,* until then a conventional review series, is entirely given over to a 435-page work by T C Cheng ' Marine molluscs as hosts for symbioses '. An examination of J P Hirth and G M Pound *Condensation and evaporation* (Pergamon, 1963) soon reveals that it is also known as volume 11 of *Progress in materials science.* But this is not really the best way to make use of this particular manifestation of the literature of science and technology. The ideal review, in the words of the preface to the second (1961) *Advances in computers,* should be ' long enough to introduce a newcomer to the field and give him the background he needs, yet short enough to be read for the mere pleasure of exploration '.

Topical surveys of this kind can obviously be published in a variety of forms, but it is the burgeoning review series such as *Advances in* chemical engineering, *Progress in optics*, *Recent progress in surface science*, *Macromolecular reviews*, which have been responsible for the spectacular rise to its present prominence of this form of scientific and technical literature. Each of these new series follows more or less the same pattern, with separate volumes containing half-a-dozen or more review articles, appearing at intervals. The titles chosen often reflect their selective, topical, flexible character, eg, *Topics in stereochemistry*, *Modern aspects of electrochemistry*, *Essays in biochemistry*, and some even show a touch of imagination, eg, *Vistas in astronomy*. Perhaps the most descriptive title is *Current topics in developmental biology*.

Freed as they are from the necessity of surveying a particular period in time, they are far less likely to appear regularly every single year: between the first and second volumes of *Currents in biochemical research* ten years elapsed; on the other hand the annual *Chromatographic reviews* changed in 1967 to twice or three times a year. With subjects chosen for their topicality *ad hoc* for each volume there can obviously be no claim that the whole field is systematically covered, but editors do take care to cast their net widely, and some even have elaborate schemes of cycling to ensure total coverage within a measureable period. For instance, *Progress in ceramic science* explains that 'within a single volume no attempt will be made to associate topics with each other, but it is hoped that in the first few volumes most of the important parts of the field will be reviewed'. The introduction in the first volume (1958) of *Advances in petroleum chemistry and refining* outlined its plan to produce ' at annual intervals a collection of progress reports written by leading authorities on particular subjects . . . In the course of several years, the series of Advances will have touched upon all parts of the far-flung industry, and the set of volumes will assume the character of a reference book. Cumulative indexes . . . will tie the material together.' By the tenth volume (1965) the grand design must have been completed, for its preface described it as ' the final volume of the series '.

Reviews need not be published in collected volumes: they can be issued separately as are the paperback Sigma science surveys, deliberately limited to 5,000 words in length and issued at the rate of four a month. Some appear in both forms: each article in the annual *Progress in materials science* is also published separately to make it available quickly, and the preface to the 1966 *Advances in applied mechanics* announced its intention ' to publish at least the next volume of "Advances" in the form of several successive fascicles, in order to present the material as rapidly as possible '.

A popular way for ' state-of-the-art ' surveys to appear is in the form of papers read at conferences, and these may be later published separately in a periodical or collected in a volume of conference proceedings (see chapter 12). Occasionally the whole conference may consist of reviews or review-type papers: the series *Advances in astronautical sciences* are the proceedings of annual and other meetings of the American Astronautical Society, and *Progress in astronautics* is based on papers read at symposia of the American Rocket Society.

But by far the oldest form of publication for a review is as an article in a regular scientific or technological journal, *eg, Analyst, Biochemical journal, Journal of chemical education, Endeavour.* Such reviews appeared in the earliest journals like the *Philosophical transactions of the Royal Society,* and a survey published in 1964 indicated that more than half of all reviews still appear in primary research journals, compared with 7 percent in annuals and other special types.[3] For many years, however (and increasingly of late), there has been a special category of periodical solely devoted to review articles, *eg, Science progress, Chemical reviews, Biological reviews, Quarterly review of biology, Contemporary physics.* Apart from their format and frequency, these review journals are often indistinguishable from the review series discussed above: the editorial policy of the *Review of modern physics* is that ' The best papers in the journal should be milestones of physics, embodying the intellectual contributions of hundreds of others whose work appears in the original literature '; in every issue of the *Quarterly reviews* (of the Chemical Society) appears the statement ' The Journal and Annual Report interest primarily the research worker: *Quarterly reviews* is designed for a wider range of readers '.

BIBLIOGRAPHICAL CONTROL

The ways in which reviews of progress serve the scientist and technologist are obvious. Currently they enable him to remain aware of the major advances outside his own particular area of activity; retrospectively he finds the bird's eye view by a perusal of a good review article is often the ideal way to orientate himself in a field compara-

---

[3] C Fix and others *Some characteristics of the review literature in eight fields of science: a report to the Office of Science Information Service* (Washington, 1964). PB 167625.

tively unfamiliar. As for the exhaustive approach, the first thing an investigator looks for is a review article or a bibliography, preferably both. Recent surveys of use have demonstrated their popularity again and again. The major survey of 3,021 physicists and chemists carried out by the Advisory Council on Scientific Policy demonstrated the almost universal use of reviews: over 90 percent of the sample had read or consulted a review within the previous month, well over half rated them as the most useful source for current awareness (a higher proportion than either abstracts or conferences) and between 46 percent and 55 percent would like more of them.[4] A more recent survey of 2,702 mechanical engineers (as a group far less literature-conscious than pure scientists) showed that reviews are used consistently by workers in all types of activity (eg, management, research, design, testing, sales, production, etc) and far more frequently than abstracts and indexes by all except those in research. Over half thought it would be useful if more review articles were published.[5]

So the demand is there. The attempt by librarians to satisfy it soon brings home the fact that even for the resources which already exist bibliographical control is inadequate. A useful aid to identify appropriate collected reviews is the UNESCO *List of annual reviews of progress in science and technology* (Paris, 1965), with some two hundred titles in subject order; the National Lending Library for Science and Technology has produced lists which include review journals as well, *eg,* *Some current review series* (1964) and *KWIC index to some of the review publications in the English language* (1966). Many of these series take particular care over indexing their contents to enable individual reviews or topics to be located: not only is it common to find each volume indexed by subject (and often by author), but regular collective indexes are frequent, *eg,* every five years for *Reports on progress in physics.* Cumulative indexes are even more useful, *eg,* to the first 60 volumes of *Chemical reviews,* and the first 46 years of *Annual reports on the progress of chemistry.* As an alternative to separately published indexes of this kind, several series include collective and cumulative indexes in the volumes themselves: an author and title index of volumes 30 to 35 is included in volume 35 (1966) of *Annual review of biochemistry;* contents lists for the preceding

[4] B H Flowers ' Survey of information needs of physicists and chemists ' *Journal of documentation* 21 1965 83-112.

[5] D N Wood and D R L Hamilton *The information requirements of mechanical engineers* (Library Association, 1967).

volumes are included in each issue of *Advances in heterocyclic chemistry;* volume 29 (1967) of *Advances in enzymology* contains a cumulative index to all the 29 volumes. But this is merely the tip of the iceberg: as we have seen, only a minority of review articles appear in such collections. The remainder, in their thousands, are scattered throughout the regular periodicals, and it is for just such a review *on a specific topic* such as waves, metal solutions, ignition and combustion of solid rocket propellants (rather than a review series) that is most commonly demanded. Of course, they are included (although not always abstracted) in the indexing and abstracting services with other periodical articles, and can be traced in the same way: it is estimated, for instance, that 6 percent of the entries in *Chemical abstracts* are for review articles. But like bibliographies (see page 94) there is no certain way to ensure that reviews on a topic emerge at the beginning of an investigation. They are embedded in the literature and it can take an exhaustive search to prise them loose. Fortunately, the need for special tools is gradually being recognised and there are now available a handful of bibliographies confined to reviews: *Bibliography of medical reviews* is an annual listing based on the corresponding section in the monthly *Index medicus; Bibliography of reviews in chemistry* derived similarly from *Chemical abstracts,* but ceased publication after 1962 for lack of support, although there are plans to revive it. The annual cumulated compilation by D A Lewis *Index of reviews in organic chemistry* (Welwyn Garden City, ICI Ltd, Plastics Division) with over 7,000 references is a lone example of its type. Norman Kharasch and others *Index to reviews, symposia volumes and monographs in organic chemistry* (Pergamon, 1962-) goes back as far as 1940 in coverage in the three volumes so far published.

It is clear that bibliographically much remains to be done. And if what we read in the introduction to *Macromolecular reviews* for 1966 is true it must be done soon: '. . . the review article is becoming the primary [*ie,* principal] source of information to a large majority of scientists '.

FURTHER READING
' Critical reviews [a symposium] ' *Journal of chemical documentation* 8 1968 231-45.

12

CONFERENCE PROCEEDINGS

As sources of information for scientists, formal meetings to hear of their colleagues' latest thoughts and discoveries have a long and respectable history. Indeed, as we have seen, it was the need first felt in the seventeenth century for a literary vehicle in which to report the proceedings of such meetings that provided some of the impetus for the earliest scientific periodicals. Today such conferences range from small gatherings of the local branch or specialist section of a professional or learned society such as the Royal Institute of Chemistry or the Institution of the Rubber Industry to the great international scientific congresses with thousands of delegates from all over the world. Most commonly described as conferences or conventions they may also be known as symposia, seminars, sessions, workshops, round tables, clinics, institutes, colloquies, or (more recently) teach-ins. Many are still announced simply as meetings. A 1965 United States estimate suggested that about an eighth of expenditure on information goes on conferences and meetings (about $50 million at that time), and an earlier survey showed that 94 percent of professional scientific societies in the US organise annual meetings at which papers on original research work are presented. If we confine our attention to international meetings, we discover that the number arranged each year amounts to thousands in science and technology alone, and the papers read at each quite frequently run into hundreds (as at the American Society of Mechanical Engineers or American Chemical Society meetings) and occasionally into thousands (2,100 at the Second International Conference on the Peaceful Uses of Atomic Energy).

Quite obviously they play an important role in scientific and techno-logical communication. Many of the papers presented often report research work several months before publication in the periodicals: indeed a feature of conferences is the large number of interim reports on incomplete work which would not otherwise see the light of day until the final report is ready—often very much later. They are clearly ' an early and important outlet in the dissemination process '. Common too as conference papers are ' state-of-the-art ' surveys, which as indicated in the previous chapter are among the most sought after of all types of sources in an increasingly turbulent sea of information.

As a medium of communication such conference papers have the great advantage of oral presentation, with questions from the audience, immediate informed criticism and comment, and follow-up contacts. Indeed, conferences can still furnish for the individual research worker many of the benefits of the person-to-person communication that was commonplace in the golden age of science when it was still possible for a scientist to know every other worker in his field. Quite truly it has been said by the investigators on the Project on Scientific Information Exchange in Psychology, which initiated the recent series of studies of the value of conferences to the scientist, that ' the convention offers, among all channels, the greatest range, both in degree and number, of opportunities for scientific communication. Considering but a single paper, an attender can choose to establish any degree of contact with its content or its authors, from merely glancing at the abstract in the programme to attending the session and approaching the authors to discuss specific questions or to pursue common scientific interests.'

And this still does not make explicit the most important function of such gatherings: in the words of a *Nature* editorial on the topic in 1967, ' the special value of conferences and symposia is that they bring people together in ways which permit informal exchanges of ideas and information '. There is evidence to show that these unplanned ' corridor ' conversations when delegate meets delegate are often for individuals quite as valuable scientifically as the formal papers when the speakers address the assembled delegates.

Of course conferences are not without their critics, who point out that much of what is heard from the speakers repeats what is elsewhere in the literature: for example, the contents of almost a third of the reports at the XVIIth International Congress of Psychology had been previously published in a scientific journal. Since for many conferences papers are not subject to any screening before presentation, the complaint of uneven quality is often heard. Occasionally one hears the charge (particularly at technological conferences) that some papers are sales-oriented. But perhaps the most serious criticism is an implied criticism, not of the conferences themselves but of their published proceedings. Conference papers are very similar in content and form to papers published in periodicals, which are the most frequently used source in most scientific and technological disciplines. Yet despite the high value placed on conferences as such by most scientists and technologists, surveys have shown low use of published proceedings. And since this is a problem of direct and particular relevance for the librarian, its investigation will take most of the remainder of this chapter.

It would be as well to start with the warning by two previous investigators that conference proceedings are ' a source of unending bibliographical confusion ', posing an ' intractable problem—the variety and complexity of the " conference " in all its maddening pre-publication and post-publication aspects '.[1] We have already seen, for instance, that conferences can range in appeal from the merely local to the worldwide, and in attendance from a handful to several thousand. They can also be ' one-off ' (eg, Conference on water utilization and effluent treatment, Edinburgh, 1969), or annual (eg, British Psychological Society annual conference, Southampton, 1970), or irregular (eg, the International congresses on polarography : 1st, Prague, 1952; 2nd, Cambridge, 1959; 3rd, Southampton, 1964, etc). The term ' international ' may mean no more than that there are delegates or speakers from more than one country. On the other hand, it may mean that the papers are in two or more languages. There may be one specialist topic for the conference, or it may be fairly general, or there may be no theme at all! It may be planned as highly scientific, for experts only, or may be deliberately designed to facilitate exchanges across disciplinary boundaries.

PRE-CONFERENCE LITERATURE
If variety is found in the conferences themselves, even more can be observed in the documents they generate. Before the conference commences, in addition to the expected announcements, calls for papers, programmes, etc, it is increasingly common to find preprints of the actual conference papers, which may give the full text, or an

[1] N J Chamberlayne and Herbert Coblans ' The proceedings of meetings: their identification and cataloguing ' *Revue internationale de la documentation* 31 1964 46-9.

abstract, or an otherwise abbreviated version. These may be duplicated (often reproduced by photolitho-offset from the authors' original type-script) or they may be printed; usually they are issued as separates, but could be in book form, *eg*, the 600 papers of *Electron microscopy: fifth international congress, Philadelphia, 1962* (Academic Press, 1962) in two volumes. They may appear in various sizes (for the same con-ference), at various times, and may be incomplete (*ie*, some papers but not all). It is not unknown to find them unnumbered and unpaged, and without the title, or date of the conference. Of course, in many cases they are not intended to have a permanent value: their aim is to act as a basis for discussion, allowing the speaker more freedom to com-ment.[2] For the librarian, however, the real difficulty is that they are usually supplied only to registered participants, and are hardly ever available to librarians in the normal way. Unfortunately, this does not prevent them being cited in the literature and asked for by a library's readers!

LITERATURE PUBLISHED DURING A CONFERENCE

This does not bulk large, being confined to texts of opening and closing addresses, lists of participants, texts of resolutions (draft and final), etc, but can be very difficult of access. Indeed, it is often impos-sible to lay hands on much of it without actually attending the con-ference.

POST-CONFERENCE LITERATURE

These are the documents most generally understood by the term ' conference proceedings ' and usually comprise the published texts of papers delivered (corrected where necessary), discussions arising, and sometimes minutes, resolutions, etc. They appear in a variety of forms:

a) *As a book: eg, Science and technology in developing countries: proceedings of an international conference, Beirut, 1967* (Cambridge UP, 1969); *The pre-Cambrian and lower palaeozoic rocks of Wales: report of a symposium, Aberystwyth, 1967* (Cardiff, University of Wales Press, 1969). Not infrequently, they extend to more than

---

[2] Conference-goers in all disciplines will confirm, however, that it is still not clearly agreed how a speaker should use his preprint. Should he assume that it has been read by the delegates, and take it from there? Should he read it word-for-word from beginning to end, even though it has been in delegates' hands for weeks and is probably before their eyes while he is speaking? Both approaches are to be found, even within the same conference.

one volume, *eg, Conference of Commonwealth Survey officers, Cambridge, 1967: report of proceedings* (HMSO, 1968) in two volumes (though the thirteen chapters are also available separately); *Genetics today: proceedings of the XIth International Congress of Genetics, The Hague, 1963* (Pergamon, 1963-5) in three volumes; *Complete proceedings of the fourth International Congress of Biochemistry, Vienna* (Pergamon, 1960) in fifteen volumes. Some of the best-known examples are issued as separate volumes in a regular series such as the Society for General Microbiology Symposia, *eg, Virus growth and variation: ninth symposium* (Cambridge UP, 1959) or the Institute of Biology Symposia, *eg, The problems of birds as pests: proceedings of a symposium, 1967* (Academic Press, 1968). Other well-known series are the Ciba Foundation Symposia, the Cold Spring Harbor Symposia, the Society for Experimental Biology Symposia.

The examples so far quoted are issued by the normal range of publishers, whether university, governmental, commercial, or other. It is quite common, however, to find the publishing undertaken (and presumably underwritten) by the sponsors of the conference, *eg, The biological effects of oil pollution on littoral communities: proceedings of a symposium, Pembroke, 1968* (Field Studies Council, 1968); *Conference on nucleonic instrumentation, Reading, 1968* (Institute of Electrical Engineers, 1968).

b) *In a periodical:* There is room here for a wide range of bibliographical permutations. As has been mentioned, the reporting of meetings has been one of the major functions of the scientific periodical for three hundred years, and it is still common to find the papers of conferences so printed as part of one of the regular issues of a journal, particularly the organ of a learned or professional society. Indeed, if the value of the paper warrants it, the whole of a particular issue may be given over to the proceedings, making it a ' conference number '. Some journals, in fact, are devoted largely if not exclusively to recording meetings, *eg, Annals of the New York Academy of Sciences, Institution of Mechanical Engineers proceedings.* Volume 182 (1967/8), part 3H of the latter, for instance, comprises (at £15) *Thermodynamics and fluid mechanics convention, Bristol, 1968.* Frequently it happens, however, that the proceedings of a particular conference are too extensive to be included within a single issue of the journal, and they have to be spread over two or more issues, sometimes extending over months or even years. And it is very common for the individual papers for a conference to be published in a number of different journals, often quite haphazardly, and not always with a clear indication that they are papers that have been presented at a particular conference.

A frequent compromise giving great flexibility is for the periodical to issue the proceedings as a supplement or a special number, additional to its regular sequence of issues, *eg, Chromosomes today: proceedings of the second Oxford chromosomes conference, 1967* (Edinburgh, Oliver and Boyd, 1969), which is a supplement to the journal *Heredity*, volume 29.

c) *In both forms:* Dual publication is not common, so far as complete proceedings are concerned, though it does occur, *eg, Biological interfaces . . . proceedings of a symposium sponsored by the New York Heart Association* (Churchill, 1968), which was simultaneously published as a supplement to the *Journal of general physiology*, volume 52, number 1, part 2, 1968. What is more frequent is the publication of individual papers in journals as well as in conference proceedings.

THE LIBRARIAN AND CONFERENCE PROCEEDINGS

As if these and other vagaries of publication were not enough, the librarian has to contend with what is probably the greatest problem—conference papers that are not published at all! In a now classic study of ' lost ' conference papers (itself read as a conference paper in 1958)[3] it was shown that of 383 papers from a sample of four US conferences 48·5 percent were never published. Of course, as was pointed out above when discussing conference preprints, not all papers are intended to have archival permanence: by definition, for instance, interim reports are designed to be superseded by final reports. Some conferences deliberately discourage publication so that participants can report and discuss more freely work that is not sufficiently advanced for wider dissemination, *eg*, the Gordon Research Conferences in the US. For some workers, too, oral delivery to a select audience might be a much preferable alternative to the hit-and-miss method of publication in a journal. But, as with preprints, this does not prevent *references* to such papers appearing in the literature, and demands for copies being made in libraries. A related problem is publication of papers in a form different from the original presentation: according to one survey only 38·2 percent of the sample were published later in

[3] Felix Liebesny ' Lost information: unpublished conference papers ' *Proceedings of the international conference on scientific information, Washington, 1958* (Washington, National Academy of Sciences—National Research Council, 1959) 475-9.

'substantially the same form as in the congress preprints and trans-actions'.

An oft-heard complaint is that proceedings often appear very late, quite commonly two years after the conference, and sometimes three, four, or even more years later. Of course, publication in book form is well-known to be slower than journal publication, and the conference editor is compelled to wait for *all* the contributions to come in before he can publish (unlike the journal editor who can always relegate long-delayed papers to the next issue). One authoritative estimate that six months should suffice is probably over-optimistic.

With the variety of publication patterns outlined above, it would be surprising if there were not acquisition problems for the librarian. The discovery that wanted proceedings are out-of-print is a common experience, particularly where the papers have been published in periodicals (and not enough extra copies printed), or where they have been published by the sponsors of the conference (who without experi-ence of the market often underestimate the demand). The difficulties would be grave enough if conferences which are regularly held adopted a consistent pattern of publication, but the librarian sometimes finds three or four different methods (or no method at all!) adopted in succeeding years. And even where the same method is adopted for each of the conferences in a series, (publication in book form, for instance) it is frequent to find that publisher is different in each case, especially with the major international conferences, held in a different country each time.

That the problem is not diminishing is witnessed by the very signifi-cant increase in the number of published conference proceedings over the last twenty years. That this is not only an absolute but a relative increase the student can quite easily gauge for himself by comparing, say, the proportion of such volumes of proceedings to total titles in classes 500 and 600 in the current *British national bibliography* volumes with similar figures from earlier years. There is no doubt that they are very frequently asked for in libraries: the National Lending Library for Science and Technology is acquiring separately-published con-ference proceedings at a rate of over four thousand a year. They make up a substantial part of the literature of science and technology. Because of their form (relatively short treatment of highly specialised topics) published conference papers serve many of the same ends as periodical articles. As we have seen their mode of presentation makes them particularly well suited as 'state-of-the-art' reports, *eg, Ideas in modern biology . . . proceedings of the XVIth International Congress of Zoology* (Garden City, NY, Natural History Press, 1965) is made up of 'reports from 19 of the world's foremost biologists in a detailed examination of the major ideas in modern biology'; *Plutonium, 1960: the proceedings of the second international conference on plutonium metallurgy, Grenoble* (Cleaver-Hume, 1961) aims to present 'a balanced picture of the state of knowledge of plutonium metallurgy in 1960'. This role is often recognised by the titles given to such works: the European Brewery Convention proceedings which appear regularly every two years used to have *Progress in brewing science* as cover-title, and the third International Conference on Semiconductors, Rochester, 1958, was published as *Advances in semiconductor science* (New York, Pergamon, 1959). In new and developing fields the publication of the proceedings of a conference can mark a watershed, serving to make an important and often unifying contribution, with the volume often remaining a standard work for some years, *eg, Proceedings of the first international conference on operational research* (EUP, 1957); *Biodeterioration of materials . . . proceedings of the first international biodeterioration symposium, Southampton, 1968* (Elsevier, 1968).

The criticisms referred to earlier of papers read at conferences apply of course with even more force to published proceedings, particularly the accusation of poor quality material. This is a grave charge, and the answer lies with the conference organisers. So long as there is no editorial check corresponding to the 'refereeing' that most learned journals impose on potential contributions, and so long as sponsors for prestige reasons feel obliged to publish in full the proceedings of 'their' conference, without applying the same criteria as they would to a monograph or treatise, complaints of this kind will continue.

A further hazard in the path of the librarian seeking to help readers in pursuit of conference proceedings is the sheer difficulty of identifying what is required. Probably the largest collection of recent conference papers in science and technology is at the National Lending Library, and yet the library staff complain that requests 'present considerable difficulty. This is partly due to the fact that borrowers often find it impossible to obtain a complete and accurate reference to the conference report required.' The reason for this is technical, but quite simple: no one knows how best to describe a conference report. Even the 1967 *Anglo-American cataloguing rules* (Library Association, 1967), which devotes two pages to the topic, has not really cracked the problem. The opinion of one of the new rules' most penetrating analysts is worth quoting: ' The appalling inconsistency and complexity of presentation of such proceedings is not fully explored in the new rules, and cataloguers will continue to find great difficulty in dealing with them. I have a feeling that this is the outermost limits of author-title cataloguing, the point at which the system becomes inapplicable.'[4] In practice, we find conferences cited by one or more of the following: the official name of the meeting, the corporate body (frequently, bodies) sponsoring, the editor of the proceedings, the subject dealt with (which may not be mentioned in the official title), the place (*eg*, Rome Conference), the authors of individual papers, etc. And if one decides to rely on the title-page for a description, as often as not one finds much of this vital data absent; and one discovers that it is a frequent habit of commercial publishers to substitute a more eye-catching title for good measure.

Even when the reader (or librarian) knows precisely what he is looking for, all is not plain sailing. To locate individual papers it is quite often necessary to turn to indexing and abstracting services. Although the position has improved since the ASLIB investigation published in 1961, which showed that of 386 papers in ten conference publications only 30 percent were abstracted and 26 percent listed in the English-language abstracting tools,[5] there are still far too many conference proceedings ignored.[6] The reader with the required volume in his hands could well be forgiven for assuming he was at the end of the trail, but some of these conference reports amount to several hundred pages and, as an earlier ASLIB survey[7] showed, 59 percent are without a subject index, 66 percent without an author index, and half of them have *no index at all*, necessitating a page-by-page search. It might well be that a similar survey today would indicate an improvement in these figures, but there are those who are prepared to argue that indexes (particularly subject indexes) to some conference proceedings do not warrant the intellectual effort required. They take the line

[4] Michael Gorman 'A-A 1967: the new catalogue rules ' *Library Association record* 70 1968 27-32.

[5] C W Hanson and Marian Janes ' Coverage by abstracting journals of conference papers ' *Journal of documentation* 17 1961 143-9.

[6] In 1965 *Biological abstracts* announced that it would no longer abstract individual papers from conference proceedings.

[7] C W Hanson and Marian Janes ' Lack of indexes in reports of conferences ' *Journal of documentation* 16 1960 65-70.

that such volumes do not make a coherent subject unit like a monograph or textbook. On the other hand, however, there are instances where it has been felt worthwhile to provide cumulated indexes, *eg*, to the 1951-63 joint computer conferences of the American Federation of Information Processing Societies. It is significant that the draft British Standard on the presentation of conference proceedings says that 'An author and contributor index and a subject index shall be included '.

BIBLIOGRAPHICAL CONTROL

In an attempt to keep pace with the ever-advancing demand for conference proceedings there have been in recent years spectacular improvements in the bibliographical tools available. Nevertheless, because of the unique way conference papers are generated, the wisest counsel for the librarian wishing to keep abreast of this category of material is to identify *in advance* the appropriate meetings and then make contact at the earliest possible moment with the sponsors, or organisers, or publishers, in order to be sure of obtaining required publications.

Fortunately, there is a wide selection of lists to help locate future conferences. Apart from general lists with which the student will be familiar, such as the annual *International congress calendar* (Brussels, Union of International Associations), there are at least four devoted to scientific and technological meetings: *Forthcoming international scientific and technical conferences* (Department of Education and Science) is British and appears twice a year; the others are American and appear quarterly, *Scientific meetings* (New York, Special Libraries Association), *World list of future international meetings: part 1, Science, technology, agriculture, medicine* (Washington, Library of Congress), and *World meetings* (Newton Centre, Mass, TMIS), the most elaborate, described as 'a two year registry of future medical, scientific, technical meetings ', in two separate parts—' United States and Canada ' and ' Outside USA and Canada '.

But locating a conference and locating its proceedings are two different problems: in some cases these lists of conferences-to-come indicate plans for the publication of proceedings, but the librarian often places more reliance in the current bibliographies at his disposal, which do not list proceedings until they have actually appeared. Many such proceedings in book form are of course listed in the regular bibliographies of current books, but there are also current lists confined to conference proceedings. Again, in addition to the general lists such as the monthly *Bibliographical current list of papers, reports and proceedings of international meetings* (Brussels, Union of International Associations), there are lists specially for science and technology, *eg*, the monthly *Directory of published proceedings: series SEMT* [*ie*, science, engineering, medicine, technology] (White Plains, NY, Inter-Dok), and the quarterly *Proceedings in print* (Mattapan, Mass). In recent years coverage of the latter has been extended to all published proceedings regardless of subject. The only British-compiled list (although of course its coverage is international) is in point of fact a library accessions list: the monthly *Index of conference proceedings received by the NLL*. It does have the very positive advantage of locating an actual copy for consultation or loan, etc.

Most of these bibliographies include not only separately published proceedings but also those which appear as part of periodicals: until recently, however, there was no index (other than the general indexing and abstracting services) to individual conference papers. This gap has now been filled by the triple monthly series of *Current index to conference papers in chemistry, . . . engineering, . . . life sciences* (New York, CCM Information Corporation). Together these attempt to index the papers delivered at the conferences announced in *World meetings* (see above).

It should be remembered that many specific subject fields within science and technology have their own conference lists, *eg*, *World calendar of forthcoming meetings: metallurgical and related fields* (Iron and Steel Institute). Many journals too make a feature of lists of meetings: these are particularly useful for local and national meetings, many of which do not get into the lists mentioned earlier. *Chemistry and industry*, for instance, not only publishes a weekly schedule but issues a six-months calendar of meetings as a supplement twice a year.

For retrospective bibliographical coverage these various lists are of limited value because most of them have only been in existence a few years. The only extensive compilation covering the whole field of science and technology is the appendix ' Periodic international congresses ' in the *World list of scientific periodicals*, which is primarily a location list without bibliographical detail and excludes 'one-off' conferences. It is true there are special subject lists, *eg*, *Bibliography of international congresses of medical sciences* (Oxford, Blackwell, 1958), but the scientist and technologist is here obliged to rely largely on the general bibliographies such as *International congresses and conferences, 1840-1937: a union list* (New York,

161

Wilson, 1938); *International congresses, 1681 to 1899: full list* (Brussels, Union of International Associations, 1960), and its supplement covering 1900 to 1919 (1964); and for more recent years the *Yearbook of international congress proceedings: bibliography of papers, reports and proceedings of international meetings held in the years 1960 to 1967* (Brussels, Union of International Associations, 1969), with its predecessor covering the years 1957 to 1959, *Bibliography of proceedings of international meetings* (1964).

FURTHER READING

Harry Baum ' Scientific and technical meeting papers: transient values or lasting contribution ' *Special libraries* 56 1965 651-3.

' Symposium volumes ' *Nature* 214 1967 46.

Paul Poindron ' Scientific conference papers and proceedings ' *UNESCO bulletin for libraries* 16 1962 113-26, 165-76.

B E Compton 'A look at conventions and what they accomplish ' *American psychologist* 21 1966 176-83.

13

RESEARCH REPORTS

Nearly twenty years ago a writer in *ASLIB proceedings* warned us that 'Report literature is fast superseding the scientific periodical and the journal of the learned society as the most important medium of scientific communication, particularly in those fields of science concerned with national defence'. As we have seen, periodicals are still with us today, though under siege, but the research report has consolidated and even extended its position as one of the most vital sources of primary information and a major rival of the journal article.

Of course, scientists and technologists have been writing reports on their work since the earliest days: what has changed over the years has been their method of communicating these reports. As has been described above, the informal letter and the society meeting only sufficed while workers were few enough to permit this personal communication; then for two-and-a-half centuries the scientific journal became the accepted method of reporting new work, supplemented by the circulation of reprints in an endeavour to retain some individual contact with fellow workers. The vast growth of science and technology has now emphasised the inadequacies of the journal (see pages 110-4), and the individual research report issued as a separate document has emerged as the most successful alternative. Its advance has gone hand in hand with the staggering increase in government spending on scientific and technological research (particularly defence expenditure), for the research report is particularly suited in a number of ways to serve as the major communication medium for 'big science'. For example, it is characteristic of government research (and research under government contract) that the scientists and technologists should be obliged to provide frequent and up-to-date reports on the progress of their work for their masters. Even where it is intended that such reports should be available to the general public, the journals are not always the most suitable vehicles. In the first place, they are too slow; secondly, the report may have too limited an interest to warrant disseminating so widely; and thirdly, journals do prefer accounts of completed work rather than work in progress. Another characteristic of government-funded research is its concentration on defence, nuclear energy, space—all topics cloaked in varying layers of secrecy. The results of such research, when communicated, must obviously have limited circulation, and the best medium so far tried is the research report. The further point is sometimes made that in some (less academic?) areas of scientific and technological activity research workers have not the time or inclination to refine and polish a report to meet the usual acceptance standards of the learned journals.

THEIR CHARACTER

As the last sentence implies, many reports are fairly rough and ready documents. Typically, they are not printed, but consist of pages of duplicated typescript (with diagrams) stapled into a cover. The vast majority are not published in any normal way: common practice is to produce no more than fifty or a hundred copies for distribution, supplemented where necessary by secondary reproduction (photocopy or microform). The text is usually in a more primitive state than a periodical article or a book, not being refereed as a rule, and sometimes not even being edited. They appear in vast numbers: according to the 1963 Weinberg report some 100,000 are issued each year in the USA alone. Described sometimes as scientific, technical, or laboratory reports, they mainly describe government-sponsored research. Private industrial research reports play a particularly important role within individual companies,[1] but by definition are not generally available: if an industrial firm wishes to make information available outside its walls it does not do so as a rule by issuing its reports, but by publishing in the open literature. There has been an increase in recent years in the issue of reports on current research by universities and colleges.

It is regularly claimed that such reports form a vital part of the research literature: ' whole technologies such as aeronautics and applied atomic energy have been built up almost exclusively on the information contained in reports '. But there is a difference of opinion among scientists and technologists: some feel that it is a waste of

[1] K E Jermy ' Handling industrial (scientific and technical) confidential report material ' *ASLIB proceedings* 18 1966 206-17.

time trying to keep abreast, since any information of lasting value will eventually find its way into the standard journals. Research reports are regarded in certain quarters as a good example of the redundancy which is widespread in the scientific communication system: in a study of over a thousand such reports produced by psychologists in 1962 it was discovered that the main content of a third of them had been published in a scientific journal by 1965, more than half of them with virtually no differences. And as has been noted, many are progress reports only, destined to be superseded in the normal course of events.

RESEARCH REPORTS AND THE LIBRARIAN

Obviously, they raise many acquisition problems for the librarian, haphazard and semi-published as they are, in limited editions, by a wide variety of bodies. He can not rely on his usual source of supply, the book trade, and is often obliged to identify and approach the issuing organisation himself. So easy is it to bring out a research report, however, that there are probably over four thousand issuing sources, ranging from two-man research teams to the largest companies and universities. The situation is further confused by the many instances where the originators of the report are not the issuing body.

A unique problem of reports is security classification. Reports that are freely available are known (rather confusingly for a librarian) as ' unclassified ', and those that are not are ' classified '. Since there are commonly various grades of security classification—in order of ascending secrecy: restricted, confidential, secret, and top secret—as well as a ' declassified ' category, the librarian often finds in practice that even classified reports may be available if he can show ' need to know '. And the librarian working with such reports may be obliged to add to his burdens by the need to control and even to ' log ' their movements.

A feature of research reports is their issue in series, characterised by a letter- or number-code.[2] Originally adopted for security reasons (so that a document could be specified without revealing either its author or its title), the practice is still common of referring to a report by its number alone, eg, FRL-TN-39; PB 155056; AD 250778. The fact that each of these numbers refers to the same document merely serves to highlight another complication. Report literature is regarded by the regular bibliographical tools as ' a minefield in which

[2] Over 13,000 codes were listed by H F Redman and L E Godfrey *Dictionary of report series codes* (New York, Special Libraries Association, 1962).

only the wary venture '. By many it is completely ignored, *eg, British technology index, Applied science and technology index;* by some it is covered partially and intermittently, *eg, Physics abstracts.*

BIBLIOGRAPHICAL CONTROL

The inadequacies of the abstracting and indexing services, the unique peculiarities of the form, and pressure from librarians and others, have in recent years so stirred the consciences of a number of governments that they have made serious attempts at rationalisation and control. After all, it is the governments that are ultimately responsible for the overwhelming majority of research reports, for many of those that they do not produce themselves are issued by non-governmental organisations under government contract. Quite rightly, governments feel that the results of this research, financed by the taxpayer, should be disseminated freely unless security is likely to be jeopardised. Some, taking their responsibilities a step further, have set up official documentation centres to collect all domestic reports and to act as issuing agency, and in the case of some of the more advanced centres, to provide bibliographical control through lists, indexes and abstracts. Some centres are limited to specific subject areas, *eg,* aerospace, nuclear energy, or to a specific clientèle, *eg,* military contractors. In some cases the larger issuing agencies have devised their own biblio-graphical systems.

US GOVERNMENT RESEARCH REPORTS

As the world's largest producer of research reports it is no more than fitting that the US government's method of control should serve as a model. The keystone of several interlocking systems is the Clearing-house for Federal Scientific and Technical Information at Springfield (Va). Described as the 'focal point for the collection, announcement, and dissemination of unclassified US government-sponsored research and development reports ', each year it adds a further 40,000 titles to its collection of 600,000. As pointed out in *The use of biological literature* by R T Bottle and H V Wyatt, ' The reports taken together represent one of the greatest collections in the world of non-confiden-tial technical information '. The key to this storehouse is the twice-monthly computer-produced abstract journal *US government research and development reports* and its companion *Index.* In addition to the usual access by author and subject, this computerised index permits reports to be traced also by corporate author, contract number, or accession/report number—a feature of particular value for this type of literature. All reports are available for purchase either in hardcopy or microfiche form, and CFSTI thus distributes over two million copies a year. Rapid alerting services are provided by *Clearinghouse announcements in science and technology,* available semi-monthly in each of 46 separate subject categories, and by the Fast Announcement Service, which highlights about 5,000 selected reports a year on ' flash sheets ' compiled in 57 subject areas. *Ad hoc* literature searches are available to order.

A number of the larger report-producing agencies have their own systems of bibliographical control in addition. The two major examples (apart from the Department of Defense, of which most of the reports are confidential) are the National Aeronautics and Space Administration and the Atomic Energy Commission. NASA produces its own twice-monthly abstract journal, *Scientific and technical aerospace reports,* covering *world-wide* report literature, and AEC is responsible for *Nuclear science abstracts,* also twice monthly, and world-wide in coverage.

And it must be remembered that over 20,000 research reports a year are printed and published and made available in conventional fashion by the US Government Printing Office, the world's largest scientific publisher. Many of these appear in long-established and well-known series, *eg*, ' US Geological Survey professional papers ', ' US Bureau of Mines information circulars ', and all are listed in the *Monthly catalog of United States government publications.*

UK GOVERNMENT RESEARCH REPORTS
Just as in the US, many such reports are published and made available through Her Majesty's Stationery Office, including what is probably the world's oldest report series, the Aeronautical Research Council *Reports and memoranda* started in 1909. These are all included in the *Monthly list of government publications.* As for the semi-published reports, until quite recently the only reliable way (outside the atomic energy field) was the do-it-yourself method of identifying from directories and other similar sources those bodies likely to issue reports in a particular field of interest and approaching them direct. The situation has now been transformed by the National Lending Library's monthly *British research and development reports,* backed up by a very large world-wide collection in hard copy or microfiche (including copies of all CFSTI reports); and by the twice-monthly *R & D abstracts*

(now available in an ' unclassified' version), which lists the reports available from the Ministry of Technology Reports Centre—covering some 80 percent of all government-sponsored research and development.

The largest single producer of reports (again excluding the Ministry of Defence) has been the UK Atomic Energy Authority. With its statutory duty to communicate to British industry whatever information will enable the various applications of atomic energy to be exploited to the full, the UKAEA has an elaborate and sophisticated bibliographical organisation. Hundreds of research papers are published annually in the open literature, but for the usual reasons mentioned above reports have been found more satisfactory in many cases. All have been listed since 1955 in the monthly *List of publications available to the public*. An excellent *Guide to UKAEA documents* by J Roland Smith (UKAEA, third edition 1963) gives a full account of the system.

An interesting and integral part of the bibliographical programme of a number of these major agencies has been the establishment in ' depository libraries' of complete collections of their unclassified reports. Most extensive have been the arrangements of the (US) AEC and the UKAEA, with depository collections all over the world. Advancing technology (mainly the shift to microfiche—so easily and cheaply reproducible) has in recent years rendered less necessary the maintenance of so many collections, and the number has been reduced.

REPORTS OF SCIENTIFIC EXPEDITIONS

Making an interesting little bibliographical category of their own are expedition reports. It commonly happens that the biological, geological, and other specimens brought back to museums, universities, and learned societies provide material for continuous study over many years. Indeed of the great expeditions of the last century much of the material still awaits research workers. The scientific results therefore may be published over a very long period as a series of research reports. Frequently the series is named after the ship, *eg*, ' Challenger' reports, ' Discovery' reports. Technically they are not periodicals, since they will come to an end someday, and they are not really books. Nevertheless, they are often entered in bibliographies of periodicals, *eg, World list of scientific periodicals*, in some bibliographies of books, *eg, British national bibliography*, and in some abstracting and indexing services, *eg, Biological abstracts*. A useful list of expeditions compiled mainly to help cataloguers in establishing correct forms of entry is Eugenie Terek *Scientific expeditions* (Jamaica, NY, Queens Borough Public Library, 1952).

FURTHER READING

R C Wright ' Report literature ' Jack Burkett and T S Morgan *Special materials in the library* (Library Association, revised edition 1964) 46-59.

Bernard Houghton *Technical information sources: a guide to patents, standards and technical reports literature* (Bingley, 1967) 78-96.

J C Hartas ' Government scientific and technical reports and their problems ' *Assistant librarian* 59 1966 54, 56-9.

J C Hartas ' Technical report literature—its nature and some problems ' Bernard Houghton *Information work today* (Bingley, 1967) 77-87.

14

PATENTS

The torrent of research reports described in the previous chapter is perhaps the most conspicuous sign for the librarian of the involvement of the modern state in science and technology, but the whole of the literature bears witness to the fact that governments are now major scientific and technological publishers. Just as now there are few facets of our lives that are not subject to some degree of official control or scrutiny, so there is scarcely a subject about which there has not been some official publication, and no category of information or form of literature unrepresented in the bibliographies of, for example, Her Majesty's Stationery Office or the US Government Printing Office. It is important for the student to realise that a so-called 'official' or 'government' publication may not be overtly official in nature. While the *Highway code* obviously is, D M Rees *Mills, mines and furnaces* (HMSO, 1969), 'an introduction to industrial archaeology in Wales', is not, and could equally well have been a commercial publication. So far as science and technology is concerned, the conventional categorisation of official publications as a class apart has far less practical application for the student of the literature than it very obviously has in the social sciences. What is significant about the following titles is not that they are all official publications but that one is a dictionary, one is a handbook, one a textbook, monograph, bibliography, research journal, abstracting journal, review journal, etc: (US) Bureau of Mines *A dictionary of mining, mineral, and related terms* (Washington, USGPO, [1968]); United Nations *Fertilizer manual* (1968); *Admiralty manual of navigation* (HMSO, revised edition 1964-); C H Gibbs-Smith *The aeroplane* (HMSO, 1969-70); Ministry of Technology *Solid-liquid separation: a review and a bibliography* (HMSO, 1966); *Tropical science; World fisheries abstracts; Agricultural science review.*

This wide embrace by governments of science and technology is a twentieth century phenomenon, but there is one area of activity that has been of critical concern to the state for several hundred years, and which has produced an extensive and unique form of scientific and technological literature. This of course is the field of inventions and the duty of the state to provide for the protection of the rights of inventors. It is thought that the earliest law for the grant of patents of invention was a decree of the Venetian Senate in 1474, although there is evidence that Venice had been making patent grants as early as 1416. The oldest system with a continuous history to the present day is the British (or English, as it was originally), which for practical purposes may be said to start from the Statute of Monopolies of 1623. Patents as a form of scientific and technological literature were born when the Patent Law Amendment Act of 1852 directed that all patents subsequently granted should be printed.[1]

Essentially, a patent is a bargain struck between the state and the inventor. The state guarantees to the inventor the sole right for a certain period of years to make, use, or sell his invention, in order that he may reap a fair reward for his labours and to encourage him to further efforts. In return for its protection, the state (*ie*, the community) obtains the invention—first of all at the market price, and then, after the expiry of the guarantee period, quite freely. More importantly for our purposes, the community obtains a detailed disclosure of the invention in the form of the patent specification—a basic and unique primary source of scientific and technological information, including over the years the first unveiling of Whittle's jet engine, Baird's television, Marconi's wireless, Dunlop's pneumatic tyre, etc.

The word 'patent' means open, and is an abbreviated way of referring to 'royal letters patent', *ie*, open (not sealed) letters addressed by the sovereign to all subjects, announcing the grant of some privilege. In the case of patents of invention (there are of course other kinds of patents) this privilege is a temporary monopoly granted to the inventor. All major countries (with and without sovereigns!) have some such system, many of them based to some extent on the UK model.

BRITISH PATENT PROCEDURE

To obtain a patent, the inventor[2] has first to describe his invention in the form of a specification (usually drafted with the aid of a patent

[1] The opportunity was grasped to print all the retrospective patents also, from Number 1 of 1617 (for 'a certain oyle to keep armor and armes from rust and kanker') to Number 14,359 of 1852.

agent) and submit it with the appropriate fee to the Patent Office in London. The British system does permit him the unique facility of filing a provisional specification if he has not the complete specification worked out: this enables him to stake his claim as soon as he has the germ of an idea, sure of provisional protection while he improves and perfects his invention. He is allowed twelve months grace to file the complete specification.

Close scrutiny by the Patent Office examiners then follows. Obviously, to be accepted the invention must be new, and the existence of a previous patent is not the only ground for rejection: any prior publication (even by the inventor himself) can disqualify. It must also be useful (although this is usually generously interpreted), not obvious to those ' skilled in the art ', not illegal (*eg*, a man-trap), and not against natural laws (*eg*, a perpetual-motion machine). In general terms it must be a ' manner of new manufacture ': this would include processes, methods of controlling and testing, as well as improvements to existing ' manufactures '. Frequently the examiners suggest modifications for discussion with the inventor, but eventually (in the case of successful applications) agreement is reached and the application is accepted. In the best of circumstances this would be a slow process : when it is realised that currently the examiners investigate well over 50,000 applications each year it is not surprising that two years or more may elapse between application and acceptance.

Up to this point no information about the specification has been disclosed to the public apart from a brief and often not very informative title announcement in the weekly *Official journal* (*patents*) at the time of application. Now, however, the 7-digit patent number is assigned (the application number has been used so far), acceptance is announced in the *Official journal*, and (about six weeks later) the specification is printed in full and put on sale. If there are no objections (from other inventors challenging novelty, for instance) the patent document is stamped with the official seal and granted. This is what comprises the royal letters patent, a legal document on parchment, to be distinguished from the patent specification, the printed description of the invention. Both are known loosely as ' the patent '.

2 Today most inventors are employed by industrial and commercial companies and produce their inventions in the course of their daily work. The British system permits companies to apply for patents, and most patents are thus company patents.

The period of protection is 16 years from the date of filing the complete specification: 4 years' protection is automatically given, but for the remaining 12 years annual renewal fees are payable on a sliding scale. Although we are mainly concerned in this chapter with patents as a source of information, their primary purpose is to serve as weapons of competition: as such they are often brandished in legal disputes. In the first instance these are referred to the Comptroller, the permanent official at the head of the Patent Office. If his arbitration fails, recourse is had to the courts; initially to the Patents Appeal Tribunal (comprising one judge of the High Court) and ultimately to the House of Lords. Accounts of leading cases appear in the regularly published *Reports of patent, design and trade mark cases*. Operating as a kind of 'technological solicitor', not only in litigation matters but at the application and examination stages, is the patent agent. A register of this specialised branch of the legal profession is maintained by the Chartered Institute of Patent Agents on behalf of the Board of Trade, and published each February.

British patent law and practice are matters of some intricacy, and in the course of a few paragraphs it has been impossible to give more than the sketchiest outline. Standard texts to consult are T Terrell *The law of patents* (Sweet and Maxwell, eleventh edition 1965) and T A Blanco White *Patents for inventions* (Stevens, third edition 1962).

THE VALUE OF PATENTS
It cannot be denied that patents are probably the most neglected of the primary sources of scientific and technological information. In the ACSP survey of the information needs of over three thousand physicists and chemists (see page 149) patents scored lowest of all as useful sources of information. Even mechanical engineers, as surveyed in 1967 (see page 149), make surprisingly little use of them: over two-thirds of the 2,072 respondents hardly ever used them, and less than 6 percent claimed to use once a month or more frequently. And yet patents are perhaps the most up-to-date form of technological literature in existence, the prime record of the progress of industry. Much of the information is not available elsewhere, and for many technologies, especially chemical, almost the whole of the very latest information is contained in patent specifications. They are of value to academic scientists also, for many contain extensive discussion of the theoretical basis of the invention and concise accounts of the 'state-of-the-art'. Mellon tells us that 'studies of chemical literature are not

complete without a search of patents '. In numbers too they make up a substantial portion of the literature, with perhaps 350,000 granted yearly out of twice that number of applications. Some it is true are ' equivalents ', *ie*, virtually identical versions published in a number of countries (for patent protection is not yet international, unlike copyright), but 60,000 are published annually in the USA alone, and 40,000 in Britain.

Reasons adduced for their comparative neglect are that the information in patents is of little value, difficult to extract, and encumbered with jargon. It is true that, unlike papers in journals, patents do not set out to be informative. They are written to define a legal monopoly, and the inventor will normally divulge no more than the Patent Office insists upon, *ie*, sufficient to enable one skilled in the particular art to carry out the process. They are frequently couched in ' patentese ', and it is perhaps not surprising that many scientists and technologists see them as stumbling blocks deliberately sited to prevent them pursuing a particular path. Yet there is no doubt that they embody a vast amount of information : there are patents extending to no more than a page, but at the other extreme we have examples like British Patent 1,108,800 with 1,319 pages and 495 sheets of drawings (an IBM computer). For a worker prepared to take his time to find his way round a typical patent (with the aid of Frank Newby's excellent guide, for example, listed at the end of this chapter), the rewards in information gleaned are substantial. Patent enthusiasts even claim ' serious advantages of patent documentation over other sources of scientific and technical information. Each patent specification provides a specific solution, resulting from an invention, for a particular technological problem, whereas an article in a journal or a research report must be laboriously sifted for what is essential and not essential, trustworthy and exaggerated, before the reader can discover the author's main idea and apply it for his own purposes. This is unnecessary in a patent specification, since it has already been done by the examiner.'[3]

So far as actually utilising the information in someone else's patent is concerned, the first point to establish is whether the patent is still in force : a large number are not renewed and therefore lapse before the full term of 16 years. General experimental work based on patented information is not normally objected to, although technically this is

[3] R P Vcerasnij, Director of the USSR Central Patent Information and Technico-Economic Research Institute.

infringement. To go any further, to incorporate in a commercial process, for instance, it is necessary to approach the patentee for a licence, for which a mutually agreed fee is usually payable. The patentee can be obliged to grant a licence if he has not worked his invention on a commercial scale within three years of the grant: probably less than half of all patented inventions actually come onto the market.

BIBLIOGRAPHICAL CONTROL

Other reasons for this disregard suggested in the survey of mechanical engineers are that ' not having access to the relevant guides to the literature, engineers do not know what exists in the patent publications ', and that ' patent literature is not generally available at local level '. Possibly librarians and information officers could play a more positive role here, for bibliographical coverage, both current and retrospective, and access to material is certainly better for patents than for a number of other primary sources of information like research reports, conference papers, theses.

To begin with, there is only one publishing outlet in each country, and its products are very strictly ordered: they are, for instance, serially numbered. The latest British Patent sequence started with 100,001 in 1916 and is currently well over the one million mark. Each week the newly-granted patents are listed in numerical order with name and subject indexes in the *Official journal.* Within seven days abstracts (called ' abridgements ' by the Patent Office) are published in pamphlet form, arranged in 25 subject groups, *eg*, metal-working, transport, organic chemistry, electric power. At approximately eight month intervals (more precisely, after every 25,000 patents), an *Index to names of applicants* is published, together with the subject and name indexes and classification schedule for each of the 25 subject groups. Retrospective searching if the patentee's name is known consists in checking back through the name indexes as follows:

    1,000,001—date: index published every 25,000

    340,001—1,000,000: index published every 20,000

    1852-1930: index published annually

    1617-1852: consolidated index.

Retrospective subject searches are best undertaken through the abridgements and their indexes. During the examination stage accepted specifications are classified according to an elaborate system comprising 8 sections, 40 divisions (in 25 groups), 405 headings and 45,000 subdivisions, and the resulting class marks are printed on the patents.

The published abridgements (which are written by the examiners) are arranged according to this scheme, and for the searcher the guides to it are the *Classification key* and the *Reference index*. Obviously to keep pace with the advance of science and technology any such scheme needs constant amendment: indeed the searcher who follows the trail back more than a few years will find himself involved in a variety of different schemes, *eg,*

   1963–date: current scheme (amended from time to time)

   1930-1963: 40 (later 44) groups, 146 classes, 271 sub-classes

   1909-1930: 146 classes, 271 sub-classes

   1855-1908: 146 classes, alphabetically arranged

   1617-1876: 103 classes, alphabetically arranged.

To help convert pre-1963 class marks into post-1963 and *vice versa,* the searcher has available the *Forward concordance* and the *Backward concordance* respectively. Another useful aid to retrospective search is the *Fifty years subject index, 1861-1910.*

Patents specifications are available for purchase at 5s each from the Patent Office Sale Branch at St Mary Cray, Kent, but copies may be consulted free of charge at the National Reference Library of Science and Invention (situated in the Patent Office building in London) and at 16 other libraries spread throughout the United Kingdom. As well as all the published abridgements, indexes, keys, etc, further special facilities for the searcher are provided by the National Reference Library: the Applications Register, the only completely up-to-date record of applications pending, and the Stages of Progress Register, giving the present status of each patent, *eg,* whether in force or lapsed, expired, licence granted, opposition to grant filed, etc. A further special index is the Name Index of Latest Acceptances which is designed to cover the gap between publication of the patent and of the printed *Index to names of applicants.* Available to all by post is the standing order service whereby subscribers are supplied on publication with all patents within any one or more sub-divisions of the classification system, and the file-list service, under which searchers can be furnished with a list showing all the patents issued in the last 50 years which have been indexed under any one of the 45,000 sub-divisions.

Extensive collections of indexes and abridgements are found at many of the provincial depository libraries, and frequently there are substantial holdings of foreign patents also. J E Wild *Patents: a brief guide to the patents collection* (Manchester Libraries Committee, second edition 1966) provides a very useful introduction to one of the largest, comprising over 5 million specifications and 30,000 bound volumes.

Supplementing the official Patent Office bibliographical tools are the regular indexing and abstracting services, many of which include patents, *eg*, *Science citation index*, *Pandex*, *Chemical abstracts*, *Computer abstracts*. A number of services concentrate on patents only, *eg*, *Footwear and leather abstracts*, *Polymer science and technology patents*. Very common are those journals which include notes or abstracts of new patents as a regular feature, *eg*, *Journal of applied chemistry*, *Production engineer*, *Modern plastics*, *Textile manufacturer*, *Electrical review*, *Metal finishing*. A feature of this field is the existence of specialist patent indexing companies, of which one of the most efficient and successful (and expensive) is Derwent Publications of London, providing an extraordinarily wide range of indexing and abstracting services. Subscribers can choose to be alerted on a country basis, *eg*, *German patents abstracts*, *Soviet inventions illustrated*, *Belgian patents report*, or on a subject basis, *eg*, *Plasdoc* (plastics), *Farmdoc* (pharmaceuticals). A variety of other aids are available, including manual code cards and computer search facilities.

PATENTS FROM OVERSEAS
The stress laid on overseas patents by a commercial service such as this reminds us how essential it is even for advanced industrial countries to keep a close watch on inventions from abroad. Because of the lack of agreement on international protection for inventions, which has already been referred to, an inventor is obliged to take out separate patents in as many countries as he feels monopoly rights will be useful. This does mean that the really important inventions get patented in all the major industrial countries—but one can never be sure! There is for instance a considerable overlap between British and US patents, but not sufficient to permit a worker in one country to ignore the patents of the other. Although starting much later than the British, US patents now far exceed them in current output and total numbers (around the 3.5 million mark). Originally modelled (like so many others) on the British system, the American patent system still operates on broadly similar lines, with bibliographical control provided by the weekly *Official gazette* with its annual indexes, and particularly since 1968 by the *Official gazette: patents abstracts section*.

A very practical reason for close observation of overseas patents is the practice in a number of countries, notably Belgium, of laying applications open to public inspection without examination for novelty very shortly after filing. Many of these are the equivalents of applications made in Britain or the US, which are not disclosed to the public until after examination and acceptance, perhaps two or three years later. The advance information thus provided has obvious commercial as well as scientific value. In a country like Belgium, the numbers involved are relatively small, but recently the burden of examining an increasing flood of applications has persuaded some other countries to follow suit. France and West Germany have changed, and Japan plans to do so soon: these are major industrial powers, responsible on their own account for large numbers of patents, but more to the point, they are countries where many British and American inventors desire protection and therefore file patent applications. In future many US and UK patents will no longer be primary sources when published.

INTERNATIONAL CO-OPERATION

A more constructive reason for concern with overseas patents is the developing interest in co-operation between countries on patent matters.[4] We have had ' convention ' patents for some years under the provisions of the International Convention for the Protection of Industrial Property: all the contracting countries undertake to grant patents without discrimination on grounds of nationality, and anyone first applying for a patent in one country can claim his date of filing as the effective date for priority of his rights in any other country in which he files a corresponding application within the following twelve months.

The most obvious success here has been the Council of Europe's International Classification of Patents[5] which has been designed to replace gradually all the national patent classifications. Comprising 8 sections, 115 classes, 607 sub-classes, and more than 46,000 groups, it is now fully operational, and from 1970 it is hoped that every published patent specification will bear an International Classification mark, as British patents have since 1957.

[4] W Weston ' Co-operation in the field of patents ' *ASLIB proceedings* 21 1969 436-43.
[5] ' The international classification of patents ' *UNESCO bulletin for libraries* 23 1969 240-4

We have yet to see an international patent, although the Nordic Patents Union (Denmark, Finland, Norway, Sweden) has made some progress, and a Europatent has been under discussion for some years. Following a study by BIRPI (United International Bureaux for the Protection of Intellectual Property), a draft Patent Co-operation Treaty has been produced which would permit an applicant fo a patent to obtain an international search on the novelty of his invention, and to select his own list of countries in which he would like protection. BIRPI has also proposed a World Patent Index.

MECHANISATION

A fruitful field for co-operation lies in the application of machines to patent searching, and much investigation has already taken place within ICIREPAT (Committee for International Co-operation in Information Retrieval among Patent Offices). Individual initiative has produced the punched card searching system at the British Patent Office and the microfilm (and aperture card) sets of US patents.

TRADE NAMES

This is probably the most convenient place to discuss trade names, for they often take the form of trade *marks*, another type of ' industrial property ' protected by legislation in a fashion similar to patents.

The simplest definition of a trade name is a name by which an article is known in commerce, such as Epsom salts, Coca-cola, Portland cement, blue vitriol, plaster of Paris, Band-aid, Leica, klaxon, etc. Of course, not all such names are proprietary: many, perhaps most, are common or generic, and for practical purposes can be regarded in the same light as, for instance, common names for plants, that is to say, as a particular kind of synonym. As such they are commonly included in encyclopedias and dictionaries of science and technology, *eg*, Arthur and Elizabeth Rose *The condensed chemical dictionary* (New York, Reinhold, seventh edition 1966), *Kingzett's chemical encyclopaedia* (Bailliere, ninth edition 1966), and *Fairchild's dictionary of textiles* (New York, 1959). There are also dictionaries, particularly in chemistry, which make a feature of their coverage of trade names, *eg*, William Gardner *Chemical synonyms and trade names* (Technical Press, sixth edition 1968), and Williams Haynes *Chemical trade names and commercial synonyms* (Van Nostrand, second edition 1955).

For hundreds of years, however, some specific trade names have been used as a device to distinguish one man's goods from another's :

indeed in these days of mass production and intensive advertising it is probable in the consumer field at least that most goods reach the market under such trade names. We do not simply buy sherry, we ask for Bristol Cream; we are urged not to be satisfied with just petrol, but to demand Shell; very few smokers are willing to purchase cigarettes irrespective of brand, for they prefer Camels or Senior Service. None of these names gives any clue as to the nature or use of the product, and yet it is difficult to see how we could do without them. They are a convenience to the consumer, and an even greater boon to the manufacturer or merchant, who, in the words of a definition of over a hundred years ago, uses them ' in order to designate the goods that he manufactures or sells, and distinguish them from those manufactured or sold by another; to the end that they may be known in the market as his and thus enable him to secure such profits as result from a reputation for superior skill, industry, or enterprise '.

Trade names of this kind are proprietary, *ie*, they belong to a particular manufacturer (or trader, etc), and in most industrial countries can be registered as trade marks[6] to obtain the protection of the law. One of the earliest (1876) and most elaborate systems is that in Britain, and so a brief description will serve as indicative of the rest. The Patent Office maintains the Trade Marks Registry, and marks are registered by a procedure similar to patenting an invention.

Applications are first scrutinised to see if they are indeed original, and the marks are then advertised (with the assigned number) each Wednesday in the weekly *Trade marks journal*. If not opposed they are registered and remain in force for seven years, though they can be renewed for further periods of fourteen years indefinitely. Marks are grouped into 34 subject classes, and registration is effective only within that class (although the same mark can be registered in more than one class if appropriate). The register is maintained in two parts : A, which gives the fullest protection and where there is no doubt that the mark is completely distinctive, and B, with a lower degree of distinctiveness, and therefore affording a less comprehensive protection.

---

[6] Of course trade names that are registered in this way make up only one of several kinds of trade mark. Because of their form they are known as ' word marks ': a trade mark may also take the form of a device, symbol, letter, signature, label, illustration, etc. Watermarks, for instance, have been known for hundreds of years; a more recent example is the British Overseas Airways Corporation ' Speedbird ', generally regarded as the best airline symbol in the world. See David Caplan and Gregory Stewart *British trade marks and symbols: a short history and a contemporary selection* (Owen, 1967).

Each year an annual name index of applicants is published, but no index of the marks themselves. However, since 1958 it has been possible to subscribe to the Patent Office service of weekly index slips of trade *names*.

There are two points worthy of mention here: the existence of a scheme for registration does not imply that unregistered marks are without legal protection. Many are quite obviously proprietary (and indeed a number date from pre-registration days): the courts will uphold the rights of owners of such marks, but the onus of proof of ownership is obviously heavier where there is no formal register to refer to. Secondly, even registration is not 100 percent watertight if the trade name is so widely used as to become generic. Proprietors of trade names need to be constantly on the alert to ensure that their name is used for their own product and that alone. Classic examples of trade names which have been lost in this way, because they have passed into language, are aspirin and bakelite. Others that have teetered on the brink for some time are Cellophane and Photostat.

There are many hundreds of thousands of trade names currently in use. As names, many of them bear no relation at all to what they represent, *eg*, a Jaguar is a car, and Stork is a margarine. They often have no meaning at all, being simply made-up words, chosen to be distinctive. Some are even computer-generated. According to one American estimate a hundred more are added to the vocabulary every month within the industrial field alone. Enquiries about trade names are very common in scientific and technological libraries, usually in the form of a request for the manufacturer of a particular branded product, or for details as to its composition and other physical properties. For the latter, the searcher has to rely largely on the encyclopedias, dictionaries, handbooks and other similar sources of such data, *eg*, the *Merck index of chemicals and drugs* (Rahway, NJ, seventh edition 1960), N E Woldman *Engineering alloys* (Chapman and Hall, fourth edition 1962), G S Brady *Materials handbook* (McGraw-Hill, ninth edition, 1963). Where the enquirer is less interested in technical information and is seeking the name and address of the manufacturer (or owner, supplier, etc), the obvious sources are the trade directories, many of which list trade names, as the index to G P Henderson *Current British directories* (CBD Research, sixth edition 1970) clearly indicates. One of the most extensive listings of US names is *Thomas' register of American manufactures*. Many trade journals make a regular feature of listing new trade names in their field.

As noted above, the official published indexes are inadequate for searches under the trade names (as opposed to searches under the names of applicants, proprietors, users, etc), but such is their importance that there are a number of commercially published search tools devoted solely to trade names. The most extensive British list is the *Kompass UK trade names,* appearing every two years with over 70,000 names in alphabetical order. Lapsed names are excluded, but it should be noted that lapsing is a matter of law, and does not mean that such a name is no longer in use. The annual *Trade marks directory* lists some 10,000. These are of course general works: there are also lists for particular industries. Two examples will suffice, both published by the appropriate trade journal: ' *The ironmonger* ' *directory of branded hardware* (over 40,000 names) and the *Food trade directory of trade marks and trade names* (over 12,000 names).

An outstanding example of a multi-purpose American list which not only gives a short description of each material and its uses as well as the manufacturer's name, but also distinguishes between registered, unregistered, and generic or common names, is O T Zimmerman and Irvin Lavine *Handbook of material trade names* (Dover, NH, Industrial Research Service, second edition 1953) and its *Supplements* (1956-65).

As with patents and other forms of ' industrial property ' there are international agreements for the protection of trade marks in the signatory countries, and as an index to such marks there is the long-established A W Metz *Intermark-index* (Zurich, 1925-   ). This work makes a brave attempt to tackle the problem of indexing non-word marks, *eg,* symbols, devices, etc, which have no obvious place in an alphabetical list.

FURTHER READING

L J H Haylor ' Scientific information and patents ' *ASLIB proceedings* 14 1962 342-9.

Bernard Houghton *Technical information sources: a guide to patents, standards and technical reports literature* (Bingley, 1967) 9-53.

Frank Newby ' Patents as a source of technical information ' Bernard Houghton *Information work today* (Bingley, 1967) 63-75.

V Tarnovsky ' Patent information services ' *ASLIB proceedings* 19 1967 332-41.

Frank Newby *How to find out about patents* (Pergamon, 1967).

Patent Office *Applying for a trade mark* (1965).

15

STANDARDS

Standards are more than a form of scientific and technological literature: without them day-to-day life as we know it would be impossible. We buy, for example, replacement electric light bulbs today without having to worry about whether they will fit our sockets. This was not always so, not until the manufacturers agreed to standardise, and British bulbs will still not fit American sockets. We can buy film for our cameras, however, in Britain or the United States, and be sure that it will fit, even though the camera is Japanese, or German, or Russian. What makes this possible is standards, national and international.

Simply and basically, standards (or standard specifications[1]) are rules as to the quality or size or shape of industrial products. This definition can be extended to include processes, methods, terminologies, etc. In a modern technological society they are essential, embracing almost every kind of scientific and technological activity. If a spanner will not fit a nut, if two cans of the same blue paint vary widely in shade, time is lost and frustration mounts. J E Holmstrom tells us: 'It has been estimated . . . that differences in the design of screw threads as between British and American practice added not less than £100,000,000 to the cost of the second world war '. And yet as long ago as 1841 in a paper to the Institution of Civil Engineers Joseph Whitworth was urging acceptance of a uniform system of screw threads.

---

[1] There is a difference between a standard and a specification, but the terms are used so loosely and interchangeably in practice (even by official bodies) that it would be fruitless as well as pedantic to attempt to distinguish them here.

TYPES OF STANDARDS

A typical standard specification is a pamphlet, no more than a few pages in length, setting out measurements, methods, definitions, properties, often with tables or diagrams. For purposes of study they can conveniently be classified according to their purpose (although there are many examples of mixed types):

a) *Dimensional standards*: these represent a deliberate effort to make things fit. They try to ensure interchangeability, so that the same products wherever and whenever they are made are identical in size, *eg*, screwdrivers and screwdriver bits for recessed head screws, crown bottle openers, WC seats (wooden) and WC seats (plastic).

b) *Performance or quality standards*: the aim here is to ensure that a product is adequate for its purpose, that it really will do what it is supposed to do, *eg*, nylon mountaineering ropes, road danger lamps, sparkguards for solid fuel fires, tarpaulins for tropical use.

c) *Standard test methods*: these help to ensure that materials and components do match up to performance or quality standards. They enable comparisons to be made on a scientific basis, *eg*, methods for the chemical analysis of ice cream, determination of stiffness of cloth, sampling and testing of paper for moisture content, method for the measurement of noise emitted by motor vehicles.

d) *Standard terminology*: by standard glossaries of terms in a particular field communication can be made more precise and accurate, *eg*, glossary of packaging terms, nomenclature of commercial timbers. Symbols too are a mode of communication, and standardisation has a role to play here, *eg*, symbols for use on flow diagrams of chemical and petroleum plant.

e) *Codes of practice*: these try to ensure correct installation, operation and maintenance, *eg*, guarding of machinery, street lighting, electrical fire alarms, frost precautions for water services.

f) *Physical and scientific standards*: these have a function different from the technical standards so far described, insofar as they deal with the physical quantities which form the basis of measurement in industry and commerce, eg, length, mass, time, temperature, etc.

Standards simplify production and distribution for the manufacturer, ensure uniformity and reliability for the consumer, and save the time of both (and of everyone in between) by eliminating unnecessary and wasteful variety, such as the classic instance unearthed some years ago by a committee of the British Standards Institution of 96 different types of garden spade in production.

The compilation of standards (and the achieving of agreement as to their implementation) is the task of industry, but it is common to find in many industrial countries a central standards organisation, usually with government support, responsible for co-ordinating effort and issuing the standards. Although there are a number of bodies in Britain which devise their own standards, eg, the Cement and Concrete Association, most work through the main organisation, the British Standards Institution, an independent body with its own Royal Charter, but supported by government grant. Responsible for over 5,000 standards, it has its own technical staff of over 200, but does its work through a series of committees representative of every part of industry. Great care is taken to consult interested bodies and individuals during the compilation of standards, and draft copies are widely circulated for comment. All the standards mentioned in the previous section of this chapter are British standards and are to be found listed (with a brief description) with all the others in the *British standards yearbook*. New standards (issued at the rate of over 400 a year) and revised standards are listed in the monthly *BSI news*. Nearly 50 subject lists (called ' Sectional industry lists ') are available free of charge, eg, Shipbuilding, Hospital equipment, Iron and steel. Particularly useful are the British standards *Handbooks* which collect together or summarise several related standards, eg, *Building materials and components, Metric standards for engineering*.

BSI operates a system of certification marking, under which by licence manufacturers are permitted to mark their goods with the BSI monogram (popularly known from its shape as the ' kite ' mark). The mark is to be seen on a wide variety of products from dustbins to clothing, and it provides an independent assurance to the purchaser that these products are produced *and tested* in accordance with the requirements of the relevant British standard. There is of course no compulsion on manufacturers to apply British or any other standards to their products: the whole system is voluntary.

At BSI headquarters a useful library and information service on standardisation is maintained, together with a complete set of British standards for reference. There is also a very large collection of standards from overseas available for loan.

UNITED STATES STANDARDS

The American standards scene offers an interesting contrast, with a very large number of bodies preparing standards, *eg*, government agencies, professional organisations, technical societies. The central organisation, the United States of America Standards Institute, is different from the BSI. It functions as a clearing-house but does not itself play the major role in preparing standards (although it may publish them). Its central function derives from its power to approve standards prepared by one of the many other bodies as ' USA standards'. Annually appears its *Catalog of standards* listing some 3,000 titles, which is supplemented by regular issues of a free list of new and revised standards. A large and comprehensive library of American and foreign standards from some 50 countries is maintained; the *Magazine of standards* is published quarterly and the *USASI reporter* twice per month.

Typical of the larger standard-preparing bodies are the American Petroleum Institute (over 100), National Electrical Manufacturers Association (over 200), Society of Motion Picture and Television Engineers (over 150), Underwriters Laboratories (200), but by far the largest is the American Society for Testing and Materials, responsible for some 4,000 standards, mainly in engineering. All of these bodies publish their own standards, whether or not they receive approval and separate publication as USA standards. The current set of ASTM standards, for instance, fills 32 volumes, now issued annually with a separately published *Index*. New and revised standards are listed in the monthly *Materials research and standards*. ASTM is obviously a major research organisation: its results are disseminated through the quarterly *Journal of materials*, and over 500 other publications in the annual *List of publications*.

The nation's central measurement standards laboratory is the National Bureau of Standards, the custodian of the basic standards of physical measurement, *eg*, mass, length, time, temperature.[2] While the bulk of industrial and other standards are prepared elsewhere, much of the basic research and testing in physics, chemistry, engineering, etc, is done at the NBS and the results published in the *Journal of research* (Sections A, B, and C), *Circulars, Handbooks, Research papers*, etc. *Publications of the NBS, 1901-47* (and supplements 1947-57, 1957-60, etc) is a comprehensive guide. Since 1969 the NBS Office of Engineering Standards Services has offered a special information service on the 16,000 published engineering standards collected from

[2] In Britain a number of similar functions are the responsibility of the National Physical Laboratory.

over 350 US trade, professional, and technical societies. And as a government agency (part of the US Department of Commerce) the NBS has a particular responsibility for helping to prepare standards for federal purchasing.

GOVERNMENT SPECIFICATIONS

Modern governments are major purchasers of many kinds of industrial products, and therefore have a particular interest in standard specifications. In many cases BSI or similar specifications seem to suffice, but commonly, the length of their purse enables them to dictate special standards of quality or performance to their suppliers: in other words they write their own specifications. These are put together with some care, often with the benefit of the best advice, and it sometimes happens that they find a wider application throughout industry. A number of them (particularly military specifications) are confidential, others are distributed in the form of unpublished documents only to those directly interested, but two major British series are available through Her Majesty's Stationery Office: the DEF specifications of the Ministry of Defence, and the DTD (Directorate of Technical Development) specifications of the Ministry of Technology. Lists appear at intervals, eg, *Aerospace material and process specifications*, DTD *series: Index* (1968); *Index of ' Defence ' publications: Part 1, Defence specifications* (1967).

The US government is the world's largest bulk purchaser, and as one might expect produces a large number of specifications. In the USA there is a much sharper division between government and industrial specifications. A useful 39-page pamphlet on this complicated pattern was produced by the US General Services Administration *Guide to specifications and standards of the federal government* (Washington, 1963). Annually published is the *Index to federal specifications, standards and handbooks*, with a monthly cumulative supplement. There is also the annual Department of Defense *Index of specifications and standards*, with cumulative supplements every two months.

COMPANY STANDARDS

Many firms have their own standards engineer or department to enable them to keep abreast of developments and to act as a link with the main standardising body, whether national, or trade, or professional. It is often the case, however, that a company is obliged to prepare its own standards, particularly when no suitable published standard exists. Sometimes these take the form of suitably modified versions of published standards, and occasionally, the positions are reversed when a private company standard forms the basis of an industry-wide or even national standard, once the need is seen to be more widespread. BSI publish a useful pamphlet on *The operation of a company standards department* (1964).

INTERNATIONAL STANDARDS

Most major industrial countries have a national system of standards. Primarily of interest to industrial users within a country's own borders, they assume greater significance as international trade grows. Manufacturers of imported goods often find it good tactics (or may even be obliged) to observe national standard specifications: conversely, exported manufactured goods may help to familiarise foreign customers with national standards. The influential position of Germany in photographic and electronic exports has made many other countries aware of DIN (Deutsche Industrie Normen) standards (for film speeds, for instance, and audio plugs and sockets). Issued by the Deutscher Normenauschuss, they are widely consulted outside Germany, and over 1,600 of them are published in an English version also, listed in *DIN English translations of German Standards* (Berlin, 1965). A valuable monthly listing of new standards from all over the world is the BSI *Oversea and Commonwealth standards*.

There have for many years been attempts at international standardisation, with varying degrees of success. The International Organisation for Standardisation, part of the United Nations Organisation, issues several hundred ISO recommendations for adoption (if they wish) by its national members, now numbering over 50, *eg*, BSI, USASI. Electrical engineering has been an area of particular international activity with both the International Electrotechnical Commission and the International Commission on the Rules for the Approval of Electrical Equipment issuing IEC and CEE specifications for international consideration. The recent change in colour coding for the cores of three-wire flexible cables and cords is on the basis of a CEE specification. A convenient listing of these and similar international standards is the *BSI yearbook supplement: Publications by international organisations*.

Like research reports, standard specifications all have a code number, *eg*, BS 3012, CP 327.404, AU 145, 2G 143 (all British standard specifications). Since it is very common for them to be cited (particularly orally) solely by their number, librarians and information officers soon learn to recognise some of the main categories by their prefixes, *eg*, NF (France), UNI (Italy), GOST (USSR), but there is as yet no one basic guide that they can turn to. The search to identify a particular code number commonly leads through the various indexes already mentioned in this chapter, bibliographies such as E J Struglia *Standards and specifications: information sources* (Detroit, Gale, 1965), dictionaries of abbreviations, and indeed any other likely source.

FURTHER READING

Bernard Houghton *Technical information sources: a guide to patents, standards and technical reports literature* (Bingley, 1967) 54-77.

Roy Binney *British standard* (Newman Neame, 1966).

A S Tayal ' Standard specifications in libraries ' *UNESCO bulletin for libraries* 15 1961 203-5.

16

TRANSLATIONS

Since the decline of Latin as the international language of scholars and scientists, communication has been increasingly impeded by the language barrier. At first, when the great bulk of work was reported in one or other of the major Western European languages (German, French, English—in that order of importance), most scientists were native speakers of one, and sufficiently familiar with the others to follow the more important discoveries. Nowadays not only is research reported in a much wider range of languages, but second only to English among the top four in terms of the amount of scientific and technological literature published is a language with which the majority of the world's scientists and technologists are quite unfamiliar, namely Russian.

English speakers in this field, as in so many others, are fortunate inasmuch as their language is more widely used than any other. In one sense this favoured position has handicapped them, and the general impression that the British (and the Americans) are not very good at languages is supported by a number of surveys. But whether they are indeed linguistic incompetents has never really been put to the test, because they have never been obliged, as say Dutch or Finnish or Czech scientists and technologists have, to learn a foreign language simply to be able to read the important literature in their own subject. Even so, 54 percent of the periodicals at the National Lending Library, 54 percent of the citations in *Index medicus* and 49·7 percent of the papers in *Chemical abstracts* are in languages other than English. About two-thirds of this non-English material is in Russian, German, and French (in that order of importance), yet an NLL survey revealed that only one scientist in ten felt he could cope with Russian, two out of three with German, and nine out of ten with French. The numbers claiming fluency were very much smaller. Over three-quarters admitted coming across a paper in the previous twelve months that they would have liked to read but could not because of the language difficulty, and almost half of them had found themselves up against this language barrier *within the previous month.*

That there is a problem, then, needs no stressing. With Urquhart many would agree that the ultimate solution lies with the working scientists themselves, and the aim should be to have within each research group one who can read French, one German, one Spanish, one Russian, one Chinese and one Japanese. In the meantime, of course, the palliative is translation. In fact, scientific and technological translation is a substantial industry, processing an estimated 4,000 million words a year in the United States alone. In the UK Patricia Millard *Directory of technical and scientific translators and services* (Crosby Lockwood, 1968) lists 300 individual approved translators, expert in 50 languages and a range of subjects from aerodynamics to zoology. A similar list for the US is F E Kayser *Translators and translations: services and sources in science and technology* (New York, Special Libraries Association, second edition 1965). In addition to freelance and part time translators, there are many in full time employment as translators in industry and commerce, government service, etc. In our subject field translating effort is concentrated for obvious reasons on primary sources, and in particular on periodical articles. Such translations may be undertaken privately by individuals; many of them are produced on a regular basis by institutions (firms, universities, learned societies, research associations, government departments, etc) for their own members and others, *eg*, the schemes operated by the UK Atomic Energy Authority, the Royal Aircraft Establishment, the Chemical Society, the Central Electricity Generating Board, the British Iron and Steel Institute, and the largest scheme of all, the US Joint Publication Research Service, a centralised service for government offices, agencies, and departments which currently produce a thousand pages of translations every working day. A number of specialist firms produce and issue translations commercially, *eg*, Henry Brutcher, Technical Translations.

But translations are expensive, sometimes as much as a hundred times the cost of acquiring the original. This is because the production of a good scientific or technological translation calls for two separate skills: thorough familiarity with *two* languages, and subject knowledge. This combination is comparatively rare, and can command an appropriate remuneration, ranging from perhaps £3 a thousand words for simple matter in the better known languages like French or Spanish to as much as £25 a thousand for turning a complex text into, say, Chinese. It is a mistake to assume that such translation is merely a matter of word-matching: exact interlingual synonyms are rare, and even where they do exist it often happens that their connotations do not match. A distinction is sometimes drawn between ' literary ' and ' non-literary ' translations, on the grounds that the specialised vocabulary of non-literary subjects makes word-matching more feasible. Many experienced translators feel that this distinction is irrational and misleading. Machine translation using the vast manipulative power of computers was once regarded as the twentieth century panacea for the woes of Babel, but despite many years of effort and tens of millions of dollars spent by the US government and others, the computer has so far made an infinitesimal contribution to the problem. Indeed, it is the opinion of a number of the world's leading authorities that machine translation is impossible. Research still continues, but the goal is not yet in prospect.

Many of the translations described above are published, or at least made available to the general public, but many more are not. Yet there is rarely anything confidential about a translation as such if the original is already available in the open literature, and many of the institutions responsible for commissioning these *ad hoc* translations have no objections to allowing others to consult them. It would be hard to think of a more fruitful field for co-operative effort, and for spreading the high cost of translations over as many users as possible. The estimated savings to the community of the National Translations Center in the USA, for instance, have been over $10 million since 1953. The first question to be asked, therefore, by the scientist or technologist faced with a paper he wishes to read in a language unfamiliar to him is not ' Who can translate this for me?' but ' Has this been translated before?'. It is to answer precisely such a question that in 1951 ASLIB established at its headquarters in London the Commonwealth Index of Unpublished Translations (with copies in other Commonwealth capitals). With over 200,000 entries, growing at a rate of 12,000 per year, this index is the obvious starting point in the search for a translation, and can be consulted without charge by post, phone, or telex. It is purely a location index of available translations : ASLIB does not hold copies of the translations, but will put enquirers in touch with the organisations which do. Over 300 sources in the UK and elsewhere co-operate by notifying new translations, mainly from Russian and German into English with the emphasis on science and technology. The success rate for consultations is of the order of 15 percent. A study some years ago showed that the estimated savings in avoidance of duplication exceeded the cost of maintaining the index by over 70 percent. ASLIB also maintains for its members a Register of Specialist Translators, recording over 200 approved names with both subject and linguistic qualifications.

TRANSLATION POOLS

One step further than the location index is the collection of translations available for consultation, photocopying or loan. Some libraries like the Science Museum Library and the National Reference Library of Science and Invention have built up extensive holdings by a positive policy of acquisition by purchase where possible, but also by encouraging the organisations or individuals responsible for commissioning translations to deposit copies in the library. In Britain one of the largest collections, with over 100,000 translations, is at the National Lending Library. Most are from Russian, and investigations here and elsewhere indicate that the demand for translations from Russian is probably equal to the total demand from all other languages. Additions to the collection donated by other organisations are listed in the monthly *NLL translations bulletin*. A feature unique to the NLL is its responsibility for the UK government sponsored translation programme. One aspect of this is the Russian Translation Scheme, under which recent articles from Russian scientific and technical journals are translated free of charge in return for a minimal share of editing of the draft by the requester. About 900 such translations a year are currently produced: they are placed on sale and announced in the *NLL translations bulletin*, and in due course are added to the loan collection. As will be seen, a commendable feature of translation pools and indexes is their close co-operation: this is instanced by the practice of ASLIB and the NLL of consulting each other before returning a negative reply to an inquiry. A further example is the inclusion by both institutions of the translations of other major translation pools, in particular the major US pools and the European Translations Centre.

In the USA responsibility is shared between the Clearing House for Federal Scientific and Technical Information and the National Translations Center. The former, a government agency (see pages 166-7), functions as a collection and distribution centre for all governmental (including government-sponsored) translations. Like the research reports handled by CFSTI these translations are announced, indexed and abstracted in *US government research and development reports* (and its Index) twice each month. The National Translations Center at the John Crerar Library in Chicago, operated by the Special Libraries Association with government support, serves as the collection centre for all non-governmental translations. Over 200 private organisations co-operate by furnishing copies of their translations, and the collection is now approaching 150,000. A *Consolidated index to translations into English* was published in 1969. New accessions appear in the twice-monthly *Translations register-index*: the translations from USGRDR and commercially available translations are also included in the index portion, which cumulates quarterly and annually.

The European Translations Centre in the Library of the Technological University of Delft in the Netherlands is both a pool and an index. Founded in 1960 under the auspices of what is now OECD (Organisation for Economic Co-operation and Development), it is a truly international exchange for some 17 co-operating national translation centres.[1] With a collection of about 150,000 translations and details of nearly three-quarters of a million in its files, ETC provides a referral service free of charge and a same-day photocopying service at cost. Its monthly bulletin *List of translations notified to ETC* includes only those *not* previously announced in national translations lists, but its quarterly *World index of scientific translations* attempts complete coverage. Computer-produced and cumulating annually, this *World index* with its arrangement in alphabetical order of the original periodical title contrasts usefully with the subject-arranged *Translations register-index*.

Many major countries have similar national pools, *eg*, the Centre National de la Recherche Scientifique in Paris with its announcement bulletin *Catalogue mensuel des traductions*; the Zentralstelle für Wissenschaftliche Literatur (Berlin), which indexes its collections in the monthly *Bibliographie deutscher Ubersetzungen*. There are also a number of international *subject* pools, *eg*, Transatom, a central information clearing house established at Brussels in 1960 by Euratom, USAEC, and UKAEA, to collect and share translations of nuclear literature, announced in *Transatom bulletin* each month.

---

[1] Of course, a translation perfectly acceptable to a polyglot Swiss scientist or Dutch technologist may not suit the British or American worker. Clearly this limits to some extent the value of this kind of international co-operation across linguistic frontiers.

The stock-in-trade of the translation pools are individual periodical articles (or conference papers, or research reports, or patents, etc), but since 1949 we have seen blossom over two hundred examples of a new kind of journal, translations from cover-to-cover of their Russian counterparts.[2] Some are commercially produced by general scientific and publishing houses, *eg, Petroleum chemistry USSR* (Pergamon) or by firms specialising in translations, *eg, Siberian chemistry journal* (Consultants Bureau), but most are sponsored by learned or professional societies, *eg, Soviet physics* (several series) by the American Institute of Physics, or by governmental agencies, *eg, Russian engineering journal* and 15 others, commissioned by the NLL from a variety of learned societies, research associations, and commercial publishers.

The advantage of such translation is that it eliminates the hit-or-miss selection of articles, and does ensure the availability (if not necessarily the use) of important foreign scientific literature. The aim, of course, is to confine cover-to-cover translation to really worthwhile journals, leaving the translation pools to cope with single articles. It is open to the criticism of wastefulness, for even the best journals must occasionally publish articles of limited value, and there are instances of articles in Russian journals that are themselves translations from the English. Naturally of course they are expensive: £50 ($120) per year is a common price for many of the commercial titles, often as much as 15 times the cost of the originals. A more severe disability is the long and perhaps inevitable delay between the appearance of the original and the translated issue: six months is common, and twelve months or more is not unknown. It has also been observed that they suffer a very high death rate. A regularly updated list is produced by the NLL: *List of periodicals from USSR and ' cover-to-cover' translations,* and a useful guide is C J Himmelsbach and G E Boyd *A guide to scientific and technical journals in translation* (New York, Special Libraries Association, 1968).

The titles chosen for cover-to-cover translation are almost invariably primary research journals, but there are a handful of secondary sources also, *eg, Cybernetics abstracts, Soviet abstracts: mechanics,* both of which are versions of the corresponding sections of *Referativnyi zhurnal;* and *Russian chemical reviews.* Not cover-to-cover translations

[2] Well over 90 percent of extant examples are translations from Russian, but there are examples from half-a-dozen other languages, *eg, Polish pharmaceutical transactions, Chinese science bulletin, Electrical engineering in Japan.*

at all, strictly speaking, but conveniently mentioned at this point are those journals containing selected translated articles gathered from a number of originals, *eg, Review of Czechoslovak medicine, Geochemistry international, International chemical engineering*; and those comprising the translated tables of contents of a predetermined selection of journals, *eg, Science periodicals from mainland China*. These categories are all included in the ETC *List of periodicals translated cover-to-cover, abstracted publications and periodicals containing selected articles*. (Delft, [1968]).

TRANSLATIONS OF BOOKS

By their nature books contain mainly secondary material and relatively few get translated: UNESCO's annual *Index translationum* is the best guide, and its arrangement (under countries) by UDC allows an approach by subject. A feature of Russian books, however, is that they do frequently contain reports of original work: the NLL has a limited programme for translating significant titles and making them available in typescript/photocopy/offset editions. These, and the additions to the large collection of Russian books, are announced monthly in the *List of books received from the USSR and translated books*.

TRANSLATIONS FROM JAPANESE

One of the countries that have grown to world stature in science and particularly technology over the last generation is Japan, and this is reflected in the literature. The translation problem that this could cause, however, is diminished to some extent by the practice of publishing many Japanese journals in English, *eg, Japanese journal of microbiology, Journal of biochemistry* (Tokyo). In science and technology as many as 12 percent appear in this way, and many of the others have English summaries or English contents pages, *eg, Seikagaku* (Biochemistry), *Konchu* (Insects). The need for an extensive programme of translation has not yet been felt, although the situation is being kept under review. Japanese translations are included in the indexes and pools, and there are a handful of cover-to-cover translations, *eg, Japanese journal of applied physics in English*.

INFORMATION ABOUT TRANSLATIONS

A major problem revealed by the NLL survey of scientists and technologists mentioned earlier is inadequate dissemination of information about the availability of translations. This chapter will have demonstrated that bibliographical control is fairly adequate: what still remains to be done is to ensure that the potential users are made aware of this. Only 17 percent of the NLL sample knew of *Technical translations* (the predecessor of *Translations register-index*) and only 7 percent used it; only 21 percent were aware of the ASLIB index and only 6 percent used it. And these, it should be remembered, were from a group of workers 76 percent of whom admitted coming up against the language barrier in the previous year. An even more remarkable revelation is provided by the list of 92 journals that the respondents suggested should be translated cover-to-cover: no fewer than 20 of them *were already available* (and listed in a variety of different sources). That ignorance of this kind is not confined to scientists and technologists can be seen from the existence side by side of two cover-to-cover translations of the same Russian original, *eg, Autometry* and *Automatic metering; Moscow University physics bulletin* and *Moscow University bulletin: physics.*

FURTHER READING

C W Hanson *The foreign language barrier in science and technology* (ASLIB, 1962).

D N Wood ' The foreign-language problem facing scientists and technologists in the United Kingdom—report of a recent survey ' *Journal of documentation* 23 1967 117-30.

J M Lufkin ' What everybody should know about translation ' *Special libraries* 60 1969 74-81.

V K Rangra 'A study of cover to cover English translations of Russian scientific and technical journals ' *Annals of library science and documentation* 15 1968 7-23.

17

TRADE LITERATURE

Of all the primary sources of scientific and technological information, probably the most neglected by librarians is the trade literature produced by industry. Issued in a tremendous variety of forms, ranging from single sheets to multi-volumed sets, by manufacturers or dealers to describe and illustrate their goods or services, such literature of course is basically advertising, and its aim is to sell a manufacturer's products or enhance his prestige. But like the advertisements in technical journals, much trade literature is far removed from the general consumer advertising familiar to us all, and is directed at a specialised audience of some sophistication. Commonly it is very technical: in the case of chemicals, for instance, as Crane points out, it will 'frequently summarize the chemistry of the compounds, give extensive information on physical properties, tell how to use them in various ways, and give references to the literature'. In very many cases it would not be untrue to say that the aim is as much to inform potential customers, users, and others, *eg*, students, teachers, research workers, about commercially available materials, equipment, and processes, as to stimulate sales as such. Increasingly too in recent years manufacturers have been stepping up the informational content of the literature to encourage users themselves to find new applications and new markets for particular products. Minute attention is often given to the production of the various leaflets and folders and brochures, not only to the actual writing (which frequently has to appeal at several technical levels at the same time), but to physical layout and production. In some companies as much as half the advertising budget is devoted to trade literature. Since 1955 we have had a British Standard Specification *Sizes of manufacturers' trade and technical literature (including recommendations for contents of catalogues)* (BS 1311: 1955), and since 1966 an *American standard for trade catalogues* (ASA z39.6—1966).

Some of these publications are merely trade *catalogues*, *ie*, basically little more than enumerations of available goods, with brief details and sometimes supplementary indexes or keys.[1] Even so they serve a vital function for the scientist and technologist: the chemist who needs a substance with certain characteristics, or the engineer looking for a piece of equipment to perform a specific task finds such catalogues invaluable, for without their aid he may not be able to ascertain easily whether they are available commercially and may thus be obliged to synthesise or build for himself. But what raises manufacturers' publications to the level of a primary information source is the continuous flow of sheets, pamphlets, bound and looseleaf volumes, on new products and processes, containing original data that has not yet appeared in the regular literature. Much of it may never be incorporated into the journals or books, and therefore some trade literature remains as a valuable supplement to the reference books and textbooks. As promotional literature it is commonly more lavishly presented and illustrated than the average professional society paper, for example, especially with graphs, tables of data, use of colour, etc. And as *industrial* literature, it is invariably firmly rooted in practice, thus forming a particularly appropriate supplement to the theory and principles of the research paper, the monograph, and the textbook.

It is being increasingly realised how valuable are retrospective collections of trade literature for studies such as industrial archeology, business history, and the history of science and technology. It has frequently been discovered that contemporary manufacturers' brochures are often the only source of information on various museum objects or industrial relics, particularly of the nineteenth and early twentieth centuries. Where such catalogues and leaflets are illustrated (as they often are) this hitherto neglected source is proving even more useful. An interesting parallel from the field of social history is the recent interest in reprinting illustrated mail order catalogues from the same period, *eg*, *Sears, Roebuck catalog for 1897* (New York, Chelsea House, 1968) and *The Army and Navy Stores catalogue, 1907* (David and Charles, 1969). And when it comes to studying the history of a firm the value of a file of their trade publications needs no pointing out. Many firms of course have had their history written, either by an independent historian (with or without the firm's blessing), *eg*, William Manchester *The house of Krupp* (Michael Joseph, 1969), or by

[1] Prices are usually omitted, although separate price lists are sometimes available on request.

someone specially commissioned for the task, *eg*, C H Wilson *The history of Unilever* (Cassell, 1954), and published through the normal book-trade channels. Such works are obviously not part of trade literature. There are, however, thousands of examples of company histories, ranging from pamphlets to multi-volumed compilations, that are not issued in this way, but are published by the companies themselves for prestige or other advertising purposes. Many are admirable, scholarly publications, of great value to the economic historian in particular, but all are perhaps best regarded as a kind of trade literature, *eg*, W & T Avery Ltd *The Avery business, 1730-1918* (1949).

It should be mentioned that not all trade literature is published by individual firms: trade associations sometimes issue catalogues listing their members' products, *eg*, *British chemicals and their manufacturers* (Association of British Chemical Manufacturers). Works of this kind are very similar in layout and use to the conventional trade directories (see pages 63-4).

FORMS OF TRADE LITERATURE
The typical piece of trade literature is a folder or pamphlet, glossily produced but commonly of an awkward non-standard size, and there is no doubt that thousands of such pieces are distributed by manufacturers daily. What distinguishes such publications from general advertising is the wealth of technical detail and the very solid body of information conveyed. Substantial pamphlets with dozens of pages of well-written text and diagrams are common, *eg*, Shell Chemicals Ltd *Building with plastics* (1965), Ferodo Ltd *Friction materials for engineers* (1961), Vinyl Products Ltd *Emulsion paint problems— causes and cures* [1966], and booklets of a hundred or more pages are frequent, *eg*, ICI Ltd Dyestuffs Division *Rubber chemicals for footwear* (Manchester, 1961). Loose-leaf format has obvious advantages for keeping such works always up-to-date, *eg*, *Colt ventilation and heating manual*.

Titles of this kind really do demonstrate how these works try to inform as well as persuade, and even more substantial examples can be found, some of them almost the equivalent of a standard work in their field, *eg*, C E A Shanahan *Chemical analysis of flat rolled steel products* (Richard Thomas and Baldwins Ltd, 1961), ICI Ltd Dyestuffs Division *The dyeing of nylon textiles* (Manchester, 1962). In some cases, they are indistinguishable from regular textbooks or monographs, save for the fact that they are issued by an industrial firm and not a publishing house, *eg*, Sir Joseph Lockwood *Flour milling* (Stockport, Henry Simon Ltd, fourth edition 1960) is the basic text on the subject. A number have attained the status of recognised reference books in their fields, *eg*, Yorkshire Engineering Supplies Ltd *Bronze: a reference book* (Leeds, [1962]), *Alcoa aluminium handbook* (Pittsburgh, Aluminium Company of America, 1962). The number of trade publications in the form of bibliographies is a further indication of the sophisticated approach to the user, *eg*, ICI Fibres Ltd *Select bibliography on nylon* (Pontypool, 1966), AEI (Manchester) Ltd *Bibliography on mass spectrometry, 1938-1957* inclusive (1961), and *Radiographic abstracts* (Ilford Ltd). The amount of literature produced by a number of the major companies is so great that some have felt it necessary to produce bibliographies of their own publications, *eg*, ICI Ltd Dyestuffs Division *Technical publications subject index up to June 1963* (Manchester, seventh edition 1964): some indication of the range of material is given by the list of series covered—sales circulars, chemicals pamphlets, technical information series, technical circulars, pattern cards, swatches, manuals.

A special form of this literature is the customer's handbook, or service manual, or user's guide, as they are variously called. These are basically textbooks and/or reference books prepared by the manufacturer for his customers on how to operate or maintain or repair his particular equipment. Perhaps the best known examples of this type are the workshop manuals for the various makes of cars, but there are similar compilations for most kinds of scientific and technological hardware, such as electron microscopes, furnaces, lathes, centrifuges, etc. Some are necessarily very elaborate, *eg*, the series of volumes known as the IBM Systems Reference Library, covering the hardware and software of all IBM computers and peripherals.

HOUSE JOURNALS

The most distinctive form of trade literature is the journal published by a particular industrial or commercial firm or public corporation, *eg, Southern electricity* (Southern Electricity Board), *Shell magazine, Atom news* (UKAEA), *Dupont magazine, Ciba review, Welder* (Murex Welding Processes Ltd). Known as house journals (house organs in USA), like the other forms of trade literature they are basically advertising publications, but in some instances they do have a great information value. In the United Kingdom the total number probably approaches two thousand, and in the United States perhaps five times that number,[2] although a large proportion are designed for internal consumption, *ie*, by the companies' own employees, and may indeed be restricted to them, *eg*, *Vickers magazine*, *The lamp* (Standard Oil Co of New Jersey), *Monsanto magazine*, *Shell news*, *Generator* (Babcock and Wilcox Co). These serve the function of newspapers within a firm, and contain information about, for instance, personnel changes, suggestion schemes, expansion plans, although a number do have roles other than communication and morale-building. The importance attached to them by the companies and the care with which they are directed at their particular audiences can be seen in the fact that the Esso Petroleum Co Ltd have a whole range of such journals: *Esso oilways*, *Esso newsline*, *Esso magazine*, *Esso news*, and *Esso farmer*.

The journals of most concern to the student are those which circulate outside the companies and these fall into three main categories:

a) *Prestige*: usually aimed at the non-technical reader, these are not necessarily concerned with a company's own products, but more with creating and preserving a favourable public image, *eg*, *Ciba journal*, *Bowater papers*, and one of the best of all, the now sadly defunct *Far and wide* (Guest Keen and Nettlefolds).

b) *Scientific/technical*: these are clearly aimed at a knowledgeable audience and qualitatively may be the equal of some of the research and technical journals described in chapter 9, *IBM journal of research and development*, *Marconi instruments*, *Hilger journal*, *Sulzer technical review*, *Machine-tool review* (Herbert Group), *X-ray focus* (Ilford Ltd), *Muirhead technique*, *Research today* (Eli Lilly & Co), *Instrument engineer* (George Kent Ltd), and one of the best known and widely respected of all, *Endeavour* (ICI Ltd).

c) *Popular*: these are similar in appeal to the commercially produced popular subject periodicals described in chapter 9 (page 109), *eg*, *Decorator news* (ICI Ltd), *Air BP* (British Petroleum Co Ltd), *Aerial* (Marconi Co Ltd), *Coal quarterly* (National Coal Board). Motoring journals are particularly well represented, *eg*, *Vauxhall motorist*, *Standard-Triumph review*, *Ford times*, *Motoring* (BLMC).

As might be expected, many house journals are fly-by-night publications, but sound, respectable and long-established titles are also found, *eg*, *Monotype recorder* (1902-), *Austin magazine* (1911-), *Osram bulletin* (1923-), *West's magazine* (1927-), *Shell aviation news* (1931-). Many are equipped with annual indexes, *eg*, *Mullard*

[2] Of course not all of these are of scientific or technological interest, *eg*, bank reviews.

technical communications, L & M news (Linotype & Machinery Ltd), *Platts bulletin,* and a cumulated index is occasionally found, *eg, Endeavour,* 1942-51.

Over six hundred titles are listed in the section ' House magazines of the United Kingdom ' in the annual *Newspaper press directory,* and they are included on a selective basis in Mary Toase *Guide to current British periodicals* (Library Association, 1962)—indicated ' H.O.', and in *Ulrich's periodical directory*—indicated ' house organ '. A directory of titles takes up most of Isobel J Haberer ' House journals ' in *Progress in library science* 1 1967 1-96. Separately published directories are British Association of Industrial Editors *British house journals* (second edition, 1962), *Gebbie house magazine directory* (Sioux City, sixth edition 1968) and ' *Printers ink* ' *directory of house organs.* A handful of the more important titles are included in indexing and abstracting services, but coverage is often selective and very patchy.

PROBLEMS OF TRADE LITERATURE

The reasons for the widespread neglect of trade literature in all save a few libraries are not far to seek, for it abounds with problems—of acquisition, arrangement, retrieval, and use.

Since virtually all such literature (including house journals) is available free of charge from the manufacturer, simply for the asking, the student might well wonder whence comes the acquisition problem. In point of fact, it is this very availability which causes one of the major difficulties: like research reports trade literature is outside the usual source of literature supply, the book trade. Booksellers are naturally reluctant to deal on a large scale with producers of literature other than regular publishers, and even more disinclined to handle free material. This means that librarians are obliged to employ direct or do-it-yourself acquisition procedures, by first identifying appropriate manufacturers from trade directories, advertisements, and other sources, and then writing either for particular items or with a request to be placed on the mailing list. Apart from the lists of house journals just mentioned, bibliographical control does not exist, for trade literature is either ignored or deliberately excluded from most current bibliographical lists (including abstracting and indexing services). The quarterly *COPNIP list* published by the Committee on Pharmaceutical Nonserial Industrial Publications of the Special Libraries Association is probably a unique example of a current list devoted to trade literature. The best sources of information on new trade publications are the scientific and technological periodicals, a number of which make a feature of noticing or at least listing new titles, *eg, Engineering, Metallurgia, R & D, Chemical week.* Advertisements also commonly indicate the availability of further information on request, and a number of periodicals provide pre-paid tear-out postcards for their readers to use in sending for such literature. Commonly each advertisement is numbered so that all that is required is for the reader to encircle the appropriate number and post the card, *eg, Chemical and engineering news, Instrument review, Modern plastics.*

But acquisition is a simple task compared to the organisation of a collection. Despite the British and American standards mentioned earlier the variety of sizes and shapes encountered is immense, and for a collection composed mainly of folders and pamphlets even the simple question of storage needs careful thought. As to arrangement and indexing, opinions differ, and this is not the place for discussion : it will suffice to indicate the nature of the problem. Ideally, any system should provide for access by name of manufacturer, name of product, trade name, and subject; yet one trade catalogue may describe hundreds, even thousands of different products. And perhaps more than any other form of scientific and technological literature the information content, and therefore its value, varies unpredictably. A particularly acute problem is maintenance: it is true that the frequent looseleaf volumes are very efficient for keeping up to date, but only at the expense of a deal of time spent in filing; and the reluctance of manufacturers to date their literature makes the task of weeding even more difficult. And because access to the information in trade literature is denied by the indexing and abstracting services, adequate arrangement, indexing and maintenance is more than usually crucial: without it a collection is virtually unusable.

COMMERCIALLY-AVAILABLE TRADE LITERATURE SERVICES
A partial solution to the librarian's problems is to subscribe to one of the ' package libraries ' or ' catalogue services ' which are now increasingly available. Known also as product information services, for an annual fee they will provide within a particular subject field an indexed collection of trade literature in standard format: the newer services will also guarantee to maintain the collection, usually on a monthly basis. A back-up enquiry service is usually available where the required item is not already in the packaged collection. The idea of

such services is far from new, but the entrepreneurial drive and efficiency of some of the recently founded examples have certainly persuaded a number of libraries of the value of trade catalogues, university libraries in particular, where previously trade catalogues had been largely ignored.

The conception of assembling in standard format the catalogues of several manufacturers goes back at least fifty years, as can be seen in the publications of organisations like the Standard Catalogue Co Ltd of London; *eg*, the five volumes of the *Architects standard catalogues, 1967-68* (seventh edition) comprise over three thousand pages on building materials, components, and services, one third of which are the manufacturers' *own* leaflets and brochures. In the United States, Sweets' Catalog Service have for many years been providing a similar bound set of manufacturers' catalogues covering a much wider field.

Perhaps because the traditional building trade is less a manufacturing industry than an industry assembling already manufactured articles (door handles, double glazing, drain pipes), its trade catalogues have always been of great importance. This accounts in part for the commercial success of perhaps the best known of the newer British trade catalogue services, Barbour Index Ltd, with over two thousand subscribers in the fields of architecture and building. More dynamic in approach than the older services, the ' package libraries ', such as this and those in the engineering field issued by Technical Indexes Ltd, are more flexible, more extensive (*eg*, 35 loose-leaf binders), and better indexed. Added to this is the monthly servicing and a regular schedule for complete replacement. The price to the subscriber is kept artificially low by persuading the manufacturers not only to standardise the format of their literature and provide it free of charge but actually to pay for having it distributed.

In 1963 the Microcard Corporation started a service to supply in the form of 4in × 6in microfiches some 14,000 catalogues of the companies listed in *Thomas' register of American manufacturers*. Use has demonstrated the aptness of microforms for rapidly changing collections of this kind, particularly now that the availability of self-threading cassette microfilm has extended the range. There are now on the market a number of new product information services using microforms, and some former ' hard-copy ' services have changed to microfiche or microfilm also. A typical system in the electromechanical field is the VSMF (Visual Search Microfilm File) of Information Handling Ltd, using some sixty 16 mm microfilm cassettes with 2,500 pages of text per cassette, and a reader-printer. Detailed indexes are provided, also on film, and updating is ensured by replacement cassettes at four-monthly intervals. Further developments in microform technology have led to the TIM (Technical Information on Microfilm Ltd) system using 6in × 4in PCMI fiches with a 200:1 reduction ratio, giving as much as three thousand pages per fiche. Initiated in late 1969, this ambitious system plans 100 percent coverage of UK manufacturers and suppliers, with complete revisions every six months, and has already been installed in the National Reference Library of Science and Invention.

These services are not without their critics: most of them are selective, and those which aim for complete coverage rarely achieve it. Not all manufacturers are willing both to supply literature and pay for the privilege (although VSMF and some other systems make no charge to manufacturers). One criticism in particular levelled at microform is its lack of colour, often so important in trade literature.

FURTHER READING

E B Smith ' Trade literature: its value, organization and exploitation ' W L Saunders *The provision and use of library and documentation services* (Pergamon, 1966) 29-54.

R A Wall ' Trade literature problems ' *Engineer* 225 1968 453-4 and 489-91.

D Kennington ' Product information services—some comparisons ' *ASLIB proceedings* 21 1969 312-6.

Jack Goodwin ' The trade literature collection of the Smithsonian Library ' *Special libraries* 57 1966 581-3.

E M Baer and Herman Skolnik ' House organs and trade publications as information sources ' American Chemical Society *Searching the chemical literature* (Washington, 1961) 127-35.

18

THESES AND RESEARCH IN PROGRESS

It is common knowledge that universities normally require a candidate for a research degree to write a thesis (sometimes called a dissertation) under the supervision of the academic staff, and that before the degree can be awarded the thesis has to be examined and approved by a recognised authority in the field. When it is recalled that in the UK two thirds of all theses are in science and technology, and that doctoral theses are usually required to show evidence of original work, it will be seen that they could make a substantial contribution to the primary literature. It is true that their main function is to allow the candidate to demonstrate his grasp both of his subject and of research method, but perhaps half of them do appear later (often condensed or amended) as articles in learned journals or conference papers or even monographs, which suggests they do have a value beyond the walls of the university.

What is not sufficiently realised is that the original theses themselves are available. Britain and America do not follow the practice of universities on the continent of Europe which normally require the candidate to have his thesis printed[1] (sometimes as many as 200 copies, which are then found very useful for exchange purposes), and typescript copies only are preserved, but these are usually available on interlibrary loan, or for microfilming or photocopying, or at the very least for consultation in the university library. Even in those cases where the research results have been published in the literature, consultation of the thesis will often bring to light background information and experimental detail, and of course the thesis is at the

[1] There are signs of a move away from compulsory printing in a number of countries in Europe, eg, Sweden. This will certainly make a doctorate cheaper for the candidate but may well mean the beginning of the end for the very valuable theses collections maintained by most university libraries in Europe, built up by exchange, and often running into tens of thousands.

disposal of investigators well in advance of publication. A large number of theses are available in microform from the specialist publishers.

Great strides have been made in recent years not only in making them more accessible but, equally important, in bringing their existence to the notice of potential users. Nevertheless, each university is still very much a law unto itself. Some publish regular lists, either separately, *eg*, the annual *Titles of dissertations approved for . . . degrees in the University of Cambridge*; or in an annual report or calendar. Others used to publish lists that are now discontinued, but many publish no list at all. Some abstracting services do attempt to include them, *eg, Chemical Abstracts, Computing reviews*; and in some fields the specialist journals may list important theses of interest, *eg*, the quarterly *NERC review* of the National Electronics Research Council. For the most part, however, the basic current bibliographies are those compiled on a national basis. France and Germany have had comprehensive official listings for nearly 90 years, but the only nation-wide bibliography in Britain is the annual ASLIB *Index to theses accepted for higher degrees in the universities of Great Britain and Ireland* which started in 1950 and now contains over 6,000 entries a year.

There is a particular difficulty experienced by users of these lists and indexes inasmuch as they provide no more than the thesis title as a guide to its content. There are grounds for believing that the titles chosen by the authors are less informative than they could be, for while they are all serious (and not eye-catching or fanciful, like the titles of some periodical articles), many are ambiguous, or misleading, or simply non-committal. The special problem this raises, of course, is that the potential user cannot just glance at the original in his library, as he can with a conference paper or journal article. With a thesis he has no alternative but to set in motion the quite expensive special machinery for obtaining access to a copy. This is why the *abstracts* of their theses that some universities publish are so particularly valuable. Since it is almost an invariable rule that candidates should provide an abstract when submitting their thesis, the suggestion has been mooted by ASLIB that these abstracts should be collected and published, at least for science and technology, as a quarterly journal.

In this respect the US has shown the way with the remarkable *Dissertation abstracts*, a monthly compilation (the separately published section B covers the sciences and engineering) of author abstracts of doctoral dissertations from over 160 co-operating universities. Developed by University Microfilms, the service will provide on demand copies (Xerox or microform) of the dissertations thus abstracted, and offers a computerised system called DATRIX (Direct Access to Reference Information, a Xerox service) for a file search of over 130,000 theses back to 1938. Not all US and Canadian doctoral theses are included, but it is claimed that they are all listed in the annual *American doctoral dissertations* which appear as issue number 13 of *Dissertation abstracts*.

So successful has been *Dissertation abstracts* that it has been decided to extend its scope to include European theses and its title to *Dissertation abstracts international*. A centre is being established at Delft to film all continental European theses and British theses will be filmed in London. Master negatives of all theses, including American, will be stored in London for reproduction on demand. What is not yet certain is how many universities will co-operate.[2]

Most of the *retrospective* bibliographies of American theses are subject lists, *eg*, M L Markworth and others *Dissertations in physics: an indexed bibliography of all doctoral theses accepted by American universities, 1861–1959* (Stanford UP, 1961), John and Halka Chronic *Bibliography of theses written for advanced degrees in geology and related sciences at universities and colleges in the United States and Canada through 1957* (Boulder, Pruett, 1958), with its continuations for 1958–63 and 1964. British theses are poorly provided with retrospective bibliographies of any kind, but Australia has *Union list of higher degree theses in Australian university libraries* (Hobart, University of Tasmania Library, 1967) and New Zealand has D L Jenkins *Union list of theses of the University of New Zealand, 1910–1954* (Wellington, New Zealand Library Association, 1956).

Master's theses present something of a problem: to begin with, they are not always required at the master's level, and in many European universities master's degrees, as such, are not granted. In research value they cannot compare with the doctoral thesis and are excluded from most bibliographies (the ASLIB *Index* is an exception). Yet they do exist, and access to them is often required. A highly selective listing appears in University Microfilms' quarterly *Master's abstracts*, but there is an attempt at a more comprehensive biblio-

[2] Gosta Ottervik and Paul Hallberg ' Microfilming and bibliographical control of European dissertations ' *Libri* 19 1969 138-41.

graphy in the annual *Master's theses in the pure and applied sciences, accepted by colleges and universities of the United States,* published by Purdue University.

RESEARCH IN PROGRESS

The requirement that a doctoral thesis must incorporate original work places on the candidate an obligation to see that his research does not duplicate what has already been done. This, of course, implies a literature search, but it also means that the candidate must try to ensure he does not choose an area that someone else has already chosen, even though he may not yet have published his results. In other words, he needs to know about research *in progress*. Of course, not only aspirants to a higher degree feel this need. Such is the pace of scientific and technological advance that this kind of research intelligence is important for all workers at what has been described as the ' cutting edge of science '. This is not merely to avoid duplication. A knowledge of research in progress in a field allied to one's own can lead to fruitful collaboration and mutual assistance. It may involve no more than an exchange of visits or of letters, but at the very least it is a point of contact and a source of information. Yet in a modern industrial state the pattern of research can be very complicated.[3] Current research and development expenditure in Britain is running at over £900 millions per year. A very large proportion (between a half and three-quarters) is supported from public funds, and, if one excepts defence research, is not too difficult to find out about. A very good guide, well-indexed, is the annual *Scientific research in British universities and colleges* in three volumes. The various bodies awarding research grants usually publish lists on a regular basis, *eg*, Ministry of Technology *Scientific research fellowships,* US Public Health Service annual *Research grants index.* Those arms of state with research programmes of their own almost always publish reports, *eg*, Medical Research Council *Current medical research,* Agricultural Research Council *Index of agricultural research,* Ministry of Technology *Research for industry: a summary of the work of the industrial research associations,* which used to appear annually, and which can be supplemented by reports from the individual associations.

[3] See the 18 reports prepared by the Organisation for Economic Cooperation and Development *Country reports on the organisation of scientific research* (*1966*).

Subject-oriented lists are occasionally encountered, *eg*, American Chemical Society *Directory of graduate research*, published biennially, which covers research in chemistry at US universities; US Public Health Service *Guide to research in air pollution* which lists current individual projects; *Space research in the UK* is produced by the Royal Society.

Considerable help is furnished by directories of scientific and technological organisations, particularly those in the form of a research guide (see pages 66-7). Access to information about research in progress in industry is less easy to come by. Such are the pressures of commercial competition that tight security measures are not unusual. Guides like *Industrial research in Britain* (Harrap, sixth edition 1968) do indicate areas of research interest of industrial companies, and approaches by genuine enquirers from outside the company are sometimes entertained.

A dream for some years has been a central index (possibly computer-based) containing details of all current scientific and technological research projects within a particular country. From this index enquiries could be answered and lists produced, as and when needed. OSTI has tried this on a pilot basis, but in the USA the Science Information Exchange has gone further. Established at the Smithsonian Institution and supported by National Science Foundation funds, it functions as a clearing-house for information on research planned or in progress, but *not* published. Data on some 100,000 projects a year is stored in a computer and an information service is provided for a nominal fee.

FURTHER READING

I R Stephens ' Searching for theses, dissertations, and unpublished data ' American Chemical Society *Searching the chemical literature* (Washington, 1961) 110-20.

Jack Plotkin ' Dissertations and interlibrary loan ' RQ 4 January 1965 5-9.

' Science Information Exchange (SIE): a national registry of research in progress ' *Scientific information notes* 1 1969 43-6.

19

NON-TEXTUAL SOURCES

Scientists and technologists are obliged to investigate many pheno-
mena that words alone are not adequate to describe, such as colours,
the songs of birds, the textures of fabrics. To read about these is not
sufficient: they must be seen, or heard, or touched. The literature, it
would seem, is no help, if by literature we mean textual sources—the
written word. Over the years, however, the writers of literature have
devised a number of parallel, auxiliary sources in an attempt to
convey information that mere text cannot. The best known of these
devices is pictorial representation (which of course actually preceded
writing chronologically) used to illustrate (literally, to light up)
textual matter. Such illustrations have amplified the information
content of printed books in science and technology since the 82
engineering woodcuts in Robertus Valturius *De re militari* ([Verona,]
Johannes Nicolai, 1472); today, advances in printing technology and
publishers' distribution methods over the last generation have permitted a larger proportion of current output than ever before to be
illustrated, often in colour. Everyone is familiar with the role of book
illustrations: there is no reason to dwell on them further. Many of
them, if not most, are helpful and convenient and agreeable rather
than absolutely essential to convey the sense of the text they illustrate. The visual material to be described in this chapter, on the
other hand, is indispensable: the information it carries cannot be
communicated by textual means.

MAPS AND ATLASES

An obvious instance is a map: to convey the information content of a
street plan of London or a contour map of the Cambrian Mountains
is virtually impossible by any other normal means. Maps (and atlases,
which are no more than volumes of maps) are widely used outside
the purely geographical field. The science of geology is largely dependent on maps: the *Geological atlas of the United States*

(Washington, US Geological Survey, 1894-1945) takes up 227 volumes. Many of the (UK) Geological Survey maps appear in parallel editions: 'drift', showing the superficial deposits, *ie*, what is actually visible, and 'solid' showing the extent and nature of the solid rocks if all the superficial deposits were removed.

In many disciplines questions of geographical distribution arise, of plants, for instance, in botany, of animals in zoology, of fossils in palaeontology, of diseases in medicine, and so maps (and atlases) become indispensable tools, *eg*, F H Perring and F M Walters *Atlas of the British flora* (Nelson, 1962), J G Bartholomew *Atlas of zoogeography* (Edinburgh, 1911), L J Wills *A palaeogeographical atlas of the British Isles* (Blackie, 1951), Royal Geographical Society *National atlas of disease mortality in the United Kingdom* (Nelson, 1963). Weather maps we are all familiar with nowadays, *eg*, J G Bartholomew and A J Herbertson *Atlas of meteorology* (Edinburgh, 1899), S S Visher *Climatic atlas of the United States* (Cambridge, Harvard UP, 1954); and the heavens too have been mapped, *eg*, A P Norton *A star atlas* (Gall and Inglis, fifteenth edition 1965), World Meteorological Organisation *International cloud atlas* (Geneva, 1956). Atlases can be found for trees, the races of mankind, ice, marine life, oil and natural gas, and many other subjects.

In *Samson Wright's Applied physiology*, a standard medical text-book (see page 78) we read: 'Anatomy is to physiology as geography is to history: it describes the scene of the action'. This is probably the explanation for the hundreds of illustrated medical works calling themselves atlases, *eg*, J C B Grant *An atlas of anatomy* (Bailliere, third edition 1951), D H Ford and J P Schade *Atlas of the human brain* (Elsevier, 1966), Perry Hudson and A P Short *Atlas of prostatic surgery* (Saunders, 1962). Such works have a long history: the first edition of the classic Spalteholz-Spanner *Atlas of human anatomy* (Butterworths, sixteenth edition 1967) appeared in Germany in 1895. A variation is the X-ray atlas, *eg*, B S Epstein *The spine: a radiological text and atlas* (Kimpton, third edition 1969); and the drug atlas, *eg*, H G Greenish and E Colin *An anatomical atlas for vegetable powders* (Churchill, 1904).

PHOTOGRAPHIC ILLUSTRATIONS
The half-tone block or some similar method of process illustration has been used in scientific and technological literature for many years. We all know that the camera cannot lie, and the accuracy of repro-

duction makes photographic illustrations (particularly in colour) of great value, for instance, in reference works designed to aid identification, *eg*, C J Morrissey *Mineral specimens* (Iliffe, 1968) is made up of over a hundred full-page coloured illustrations; similarly B J Rendle *World timbers* (Benn, 1969-   ) when completed will comprise three volumes of plates in full colour.

Of even greater use are those photographic illustrations that provide information that no other form can, such as microphotographs, *eg*, British Leather Manufacturers' Research Association *Hides, skins and leather under the microscope* (Egham, [1957]), with almost a thousand illustrations; A B Wildman *Microscopy of animal textile fibres* (Leeds, Wool Industries Research Association, 1954). Rather rarer are stereoscopic photographs, which show objects three-dimensionally when viewed with the apparatus supplied, *eg*, F C Blodi and Lee Allen *Stereoscopic manual of the ocular fundus in local and systemic disease* (Kimpton, 1964).

SPECIAL TYPES OF VISUAL MATERIAL

Workers in different subjects have developed different methods of presenting their material visually, and this has produced a whole range of non-textual information sources. A selection only can be described here. Without *circuit diagrams* the electronics engineer would find his task virtually impossible, *eg*, G A French *Twenty suggested circuits* (Data, 1960). The biologist uses *metabolic charts* to indicate the interrelations and correlations of biochemical reactions in metabolic sequences, *eg*, W W Umbreit *Metabolic maps* (Minneapolis, Burgess, 1952) and *Supplement* (1960). *Silhouettes* have been successfully used for many years to aid identification of planes and ships, *eg*, E C Talbot-Booth and D G Greenman *Ship identification* (Allan, 1968-   ). Chemists use *ring formulae* as the most convenient way of showing the molecular structure of organic compounds, *eg*, A M Patterson and others *The ring index* (Washington, American Chemical Society, second edition 1960) and *Supplements* (1963-   ). A unique work is Gerhart Tschorn *Spark atlas of steels* (Pergamon, 1963), comprising over a hundred spark photographs used as a simple means of identification, based on the fact that different steels produce different sparks when at the grinding wheel.

AUDIO-VISUAL MATERIALS

In the form of tape and disc recordings, films, filmstrips, and film loops, these are very familiar, particularly in the educational world. Less well-known as yet are electronic video recordings (EVR), but they seem destined to play a revolutionary role in this field over the next decade.

In science and technology sound recordings are not unknown, but have limited scope, *eg,* Myles North and Eric Simms *Witherby's sound-guide to British birds* (1969) comprises a book and two long-playing records. Film, on the other hand, is widespread. Filmstrips (comprising a succession of still photographs) can convey little more information than a good illustration; cine-film on the other hand adds another dimension. The remarkable cart-wheeling method of getting from place to place that *Hydra* uses, as well as its less spectacular looping and shuffling movements, can only be fully appreciated by actually seeing it, preferably *in vivo,* but if not, on film. Refinements like slow-motion or time-lapse photography extend the range of applications in science and technology of cine and video even further.

Students should be familiar with the major bibliographical aids to such material: the bi-monthly *British national film catalogue* with its arrangement by the Dewey Decimal Classification contrasts usefully with the quarterly Library of Congress *Catalog: motion pictures and filmstrips* which is arranged alphabetically by title with a subject index. Both are cumulated annually. There are a variety of other guides and indexes in specific subject fields, *eg,* Royal Institute of Chemistry *Index of chemistry films* (fifth edition 1967) with over 2,000 titles of films, filmstrips, and film loops; United Kingdom Atomic Energy Authority *Film catalogue* (1968); Ministry of Agriculture *Agricultural and horticultural films held in the film loan library* (1964). The Scientific Film Association issues a *British film guide* in several sections, each devoted to a particular subject, *eg,* medicine, biology, chemistry. Some lists include books and papers as well as films, *eg,* Building Research Station *Information 68: a list of publications and films* (1968).

The overwhelming majority of films in such lists are for teaching purposes: it could be said that they correspond to the textbook in the published literature, *eg,* 'Modern methods of underground pipe-laying' (1963), 'The fuel cell' (1956), 'The story of basic slag' (1962). Some of them are quite technical and at an advanced level, *eg,* 'Welding by tape' (1964), 'Computer graphics' (1967), and occasionally films of a research nature are found, particularly in cases where cine-film is thought to be a particularly appropriate medium for expounding the topic, *eg,* 'Air flow round buildings' (1967).

A comparative newcomer to the ranks of audio-visual teaching aids is the ' programme '. With programmed textbooks (see page 82), both linear and branching, most libraries are now familiar, but programmes are also produced in the form of film or sheets or paper rolls for use in the various types of teaching machine. A convenient general listing is the Association for Programmed Learning *Programmes in print* (1966); the thousand programmes in the British Association for Commercial and Industrial Education *Register of programmed instruction* (1968) are limited to science, technology and business studies. Examples are ' Coordinate geometry for calculus ' (1966), ' Probability statistics for managers ' (1966).

MICROFORMS

Perhaps because they too require special equipment for their utilisation, microforms are frequently grouped with audio-visual materials. Technically, they differ only in *format* from the other categories of information described in this book, and comprise textual as well as non-textual sources, but it is convenient to consider them here despite their pervasive nature.

The student will be familiar with the various kinds of microforms[1] and their uses: all that can be done here is to indicate their particular contribution as an alternative form of publication to the literature of science and technology. Their frequent appearance throughout this work may have been noticed, for among the bibliographies of microforms can be found examples of practically all categories of information sources. For very many years now they have been used for *retrospective micropublication, ie,* the reprinting of originals that for one reason or another do not warrant conventional full-size republication. Sometimes the originals are old and rare, as in the very ambitious Landmarks of Science series, which plans to issue in Microprint form the outstanding works of more than 3,000 scientists, but more often they are out-of-print works of all kinds for which there is still a demand—but not such as to justify the expense of a new edition. Indeed so extensive is the range now that it would almost be possible to build up a respectable working library of science and technology in microforms alone. By way of illustration, all the following works, chosen at random from the examples given in previous chapters, are available in microform: American Chemical

[1] B J S Williams ' Microforms in information retrieval and communication systems ' *ASLIB proceedings* 19 1967 223-31 is a convenient account.

Society monographs, *Nuclear science abstracts*, the US Joint Publications Research Service translations, *Report on the scientific results of the voyage of HMS Challenger . . . 1872-76*, the United Kingdom Atomic Energy Authority unclassified reports, *Nature, Endeavour, Shoe and leather news, Engineering, Science, American machinist*, US chemical patents, the Aeronautical Research Council *Reports and memoranda*, the Iron and Steel Institute Bibliographical Series, Dean *A bibliography of fishes*, Royal Society *Catalogue of scientific papers*, US Patent Office *Official gazette*. There are thousands of others. Not all of them are replacements for out-of-print originals, for in some cases the microform edition is on sale as an alternative to the hardcopy original, for those who may prefer it.

What has increased greatly in recent years has been *original micropublication, ie*, of material not published in any other format. Both the technology and the economics of microreproduction permit very small editions, or even one-off, on-demand publishing, which is ideal for those categories of scientific and technological literature with a limited appeal, *eg*, university theses, research reports, etc. The services of *Dissertation abstracts* represent the best known instance of thesis micropublication, and the enormous US government programme (through the Clearing House for Federal Scientific and Technical Information) is certainly the largest in the field of report micropublication. Other kinds of material for which micropublication has been found appropriate are a scientist's collected works, *eg*, Albert Einstein *Complete works* (Readex Microprint, 1966); a technologist's manuscripts, *eg*, *The papers of James Nasmyth, 1808-90* (Micro Methods, 1968); illuminated manuscripts (*ie*, in colour), *eg*, the tenth century *Bodley Herbal* (Micro Methods, 1964).

Bibliographical control of microforms has been something of a nightmare,[2] because of the comparative ease with which microform editions can be produced, but we now have the cumulative Library of Congress *National register of microform masters*. The most comprehensive trade listing, with over 15,000 entries is the annual *Guide to microforms in print* (Washington, NCR Microcard Editions) and its biennial companion *Subject guide to microforms in print*.

An even more recent feature of microforms in science and technology has been the emergence of microform ' systems ', based on the miniaturisation of a complete collection rather than individual

[2] ' Bibliographical control of microcopies ' *UNESCO bulletin for libraries* **19** 1965 136-60, 172.

documents.[3] Some examples of these 'package' or 'desk-top' libraries in the field of trade literature were quoted in chapter 17, but there are subject-oriented systems also. One of the best known is *Chemical abstracts* on microfilm—over 50 years of abstracts totalling more than 4 million housed in about 150 cassettes of 16 mm film.

THREE-DIMENSIONAL MATERIAL

There are limits even to what non-textual literature can communicate, as the rapid developments in molecular biology over the last decade have highlighted. Workers in this field concern themselves with investigating the dimensional structure of very complex molecules—their most remarkable achievement was the discovery of the structure of DNA—and make great use of three-dimensional models. Their problem is communicating with each other about molecular structures *without* the aid of models. This is but one recent instance of a larger problem, and no doubt some solution will be found, possibly *via* the computer. Valiant attempts have been made with similar difficulties in the past by providing actual specimens to accompany literature in conventional form, *eg*, Alfred Schwankl *What wood is that?* (Thames and Hudson, 1956) comprises a volume of text and 40 actual timber samples in a case; the British Colour Council *Dictionary of colours for interior decoration* (1949) is made up of one volume of text and two volumes of strips painted in matt and gloss paint and samples of pile fabric dyed in 319 colours with 6 shades of intensity.

[3] B J S Williams ' Recent developments in microform systems ' *The information scientist* 3 1969 31-8.

20

BIOGRAPHICAL SOURCES

It is no more possible in science than in any other area of intellectual endeavour to separate a man's life from his work. So intimately bound up with his personal life is the work of a research scientist that a full understanding of his achievements often demands close study of his biography. Perhaps this is less true of the technologist, and even of the scientist in many humble and mundane circumstances, but there is no doubt that biographical information is frequently sought in libraries about scientists and technologists of all kinds. The deep enquiry into the personal background of an individual corresponds to the 'exhaustive' approach to the literature described in chapter 1 (page 18). Far more common are the 'everyday' biographical enquiries, where the searcher needs either one specific fact, *eg*, the full name, or the address, or the present post of the man he is interested in; or in some cases rather more detail to enable him to come to some kind of judgment as to his status, *eg*, academic qualifications, previous experience, age, publications, etc.

The literature resources that a searcher has at his disposal are not in fact scientific and technological information sources at all, as described in chapter 1. They are *biographical* sources, and even where they specialise in scientists and technologists, they are not part of the literature of science and technology. This is a point of major significance for the searcher, particularly if he is more used to investigating scientific and technological sources of information. Because of the nature of the literature, *ie*, because it is about people rather than about science and technology, searches in the biographical sources have their own rules: it is far more common, for instance, for a biographical search to end in failure. In this science is no different from any other subject: most people are not renowned; the published biographical sources concentrate on the renowned: therefore, there is no published information available about most people. There is, on the other hand, an overwhelming amount of detail about the important figures, repeated time and again in the various sources. On the less famous, the best sources of biographical data are the directories of individual scientists and technologists, as described in chapter 6 (pages 64-6). Within their limits these are non-selective, inasmuch as a society membership list, for instance, will include all members in good standing. For the simpler everyday enquiries as to name, location, degrees, etc, such lists will often serve, but very few include real biographical information on matters such as place of birth, education, previous posts held, etc. For these one turns to the overtly biographical sources, particularly the biographical dictionaries.[1]

BIOGRAPHICAL DICTIONARIES

It is frequently drawn to our attention that over 90 percent of all the scientists who have ever existed are alive today: not surprisingly, therefore, there is a wide range of publications providing biographical information about them. Mostly of the ' who's who ' type, they are obviously selective, but do aim to include as large a proportion of the scientific and technological community as is feasible. An example of international scope is the three-volumed *Who's who in science in Europe* (Hodgson, 1967), listing 30,000 names from over 20 countries. As is commonly the case with such publications, the entries are based on the responses made by the scientists themselves to questionnaires circulated by the publishers. The word ' science ' in the title includes technology and medicine, as it also does in a contrasting dictionary of similar international scope but very different approach: *McGraw Hill modern men of science* (1966-8). Here attention is concentrated on ' leading contemporary scientists ', amounting to no more than 850 in the two volumes.

Such works are frequently compiled on a national basis: by far the largest is *American men of science* (New York, Bowker, eleventh edition 1965-) in several volumes. An unusual example of a list compiled in English but covering foreign scientists is John Turkevich *Soviet men of science* (Princeton, NJ, Van Nostrand, 1963), describing 400 of the academicians and corresponding members of the USSR Academy of Sciences. By contrast, Ina Telberg *Who's who in Soviet*

[1] Some dictionaries and encyclopedias of science and technology contain biographical entries, *eg, Hackh's Chemical dictionary,* H J Gray *Dictionary of physics.* As a rule, however, they concentrate on the famous figures.

science and technology (New York, Telberg, second edition 1964) contains translations of about 1,000 entries selected from a biographical dictionary published in Moscow.

'Who's who' compilations are commonly found in particular subject fields: *Who's who in atoms* (Harrap, fourth edition 1965) has over 20,000 entries from more than 70 countries: *Who's who in engineering* (New York, Lewis) has a similar number of entries but mainly for US engineers. Lists confined to one country are *Who's who in space* (Washington, Space Publications, second edition 1968), *Who's who of British engineers* (Maclaren), and *Electrical who's who* (Iliffe).

A feature of biographical dictionaries in many subjects is the division found between those listing living persons and those which only include names after death. All the titles so far mentioned in this chapter are of the first kind: a recent example of the second is T I Williams *A biographical dictionary of scientists* (Black, 1969), which ranges from the earliest times to the present, includes technologists as well as scientists, and covers all countries. By way of contrast the 3,000 entries in the two volumes by E G R Taylor *The mathematical practitioners of Tudor and Stuart England* (CUP, 1954) and *The mathematical practitioners of Hanoverian England* (1966) are confined to a particular period, a particular discipline, and a particular country. Not surprisingly, perhaps, many compilers and publishers find this dichotomy in print between the living and the dead to be too restricting, and some major biographical dictionaries ignore the distinction, *eg, World's who's who in science* (Chicago, Marquis, 1968), the largest one-volume compilation with 32,000 names from 1700 BC to 1968 AD; *Chambers's Dictionary of scientists* (1951) a much more selective list of 1,400 leading figures; and the great scholarly standard work, *Poggendorffs Biographisch-literarisches Handwörtebuch* (Leipzig, Barth, 1863-   ). Within particular subject fields too, it is usual to find past and present figures intermingled, *eg,* J H Barnhart *Biographical notes on botanists* (Boston, Hall, 1965), comprising in three volumes reproductions of over 40,000 cards from a file maintained by the New York Botanical Garden Library.

The well-nigh universal arrangement of such biographical dictionaries is alphabetically by biographee. It is likely that this suits the majority of searchers who set out with a name. Other arrangements are occasionally found, however, as in Isaac Asimov *Biographical encyclopedia of science and technology* (Allen and Unwin, 1966), which is subtitled ' the living stories of more than 1000 great scientists from the age of Greece to the space age chronologically arranged '. More common are supplementary indexes to a main alphabetical sequence to permit an alternative approach: probably the most extensive example is *Who knows—and what* (Chicago, Marquis, revised edition 1954), ' keying 12 selected knowers to 35,000 subjects '. Similar subject approaches are provided on a smaller scale by the indexes in *McGraw-Hill Modern men of science, Chambers's Dictionary of scientists,* and many others. Less frequently encountered is the geographical index, *eg,* in W S Downs and Williams Haynes *Chemical who's who* (New York, Lewis, fourth edition 1956).

INDIVIDUAL AND COLLECTED BIOGRAPHIES

For those scientists and technologists who have made a substantial mark there is always the possibility that they may have had a whole book devoted to them, *eg,* D McKie *Antoine Lavoisier* (Constable, 1952), R Pilkington *Robert Boyle* (Murray, 1959), F E Manuel *A portrait of Isaac Newton* (OUP, 1968), S R Ranganathan *Ramanujan: the man and the mathematician* (Asia Publishing House, 1967), and some of course have written their own stories, *eg,* Peter Scott *The eye of the wind* (Brockhampton Press, revised edition 1968). Tracing these is no problem for they appear in the standard bibliographies and library catalogues. The British Museum *General catalogue of printed books* in particular is a convenient source because of its practice of listing books about an individual under his name as well as under the author. At least one bibliography of such works has appeared: T J Higgins *Biographies of engineers and scientists* (Chicago, Illinois Institute of Technology, 1949).

Many more scientists and technologists figure as the subjects of individual chapters in works of collected biography. Samuel Smiles *Lives of the engineers* (1861-2) is a classic example, but there are countless others, *eg,* E T Bell *Men of mathematics* (Penguin, 1953) in two volumes; A Findlay and W H Mills *British chemists* (Chemical Society, 1947); L C Miall *The early naturalists* (Macmillan, 1912). In some cases such collected works are not original contributions but take the convenient form of reprints or translations of articles or extracts previously published elsewhere, *eg,* E Farber *Great chemists* (Wiley, 1961). The tracing of the accounts of individual workers in such volumes is made easier by the existence of a number of invaluable analytical compilations such as N O Ireland *Index to scientists of the world from ancient to modern times* (Boston, Faxon, 1962) which covers nearly 7,500 scientists and technologists in 338 collections. An interesting example in a specific subject field is J Britten and G S Boulger *A biographical index of deceased British and Irish botanists* (Taylor and Francis, second edition 1931).

Journals too are often valuable sources of biographical information. As well as overtly biographical articles, they often feature short biographical notes about their contributors, *eg, Vacuum.* More useful still are obituary notices: outstanding here are the series that used to appear in the *Proceedings of the Royal Society.* So valuable a source were they that since 1932 they have been published separately, and appear currently as the annual *Biographical memoirs of Fellows of the Royal Society.* A similar series is the US National Academy of Sciences *Biographical memoirs* which started in 1877. Both these major series have indexes to speed the location of an individual obituary, *eg,* the index to volumes 1-35 of the latter appeared in volume 36 (1962). Some of the better journals have cumulated indexes, *eg,* the American Society of Mechanical Engineers *Transactions* index covering 1880-1923 lists some 1,400 obituaries, but with E S Ferguson *Bibliography of the history of technology* (MIT Press, 1968) ' One is inclined merely to wag one's head sadly at the voluminous but largely unrecoverable biographical information in periodicals '.

PARALLEL SOURCES

For the major figures there are sometimes other sources of information productive of biographical details that are not themselves primarily biographical, *eg,*

a) *Correspondence:* individuals vary in the amount of self-revelation they indulge in in their letters, but it is common for published editions to include background biographical details as editorial matter, in an introduction, for instance, or in the notes, *eg,* H W Turnbull *The correspondence of Isaac Newton* (CUP, 1958-). In any case, correspondence almost invariably sheds some illumination on a scientist's work.

b) *Collected works:* these are even less patently biographical, but the very fact of gathering a man's books and papers together gives them some biographical interest immediately. Like good editions of correspondence, they often include information of an introductory or explanatory nature on the author's life, *eg, The collected papers of Lord Rutherford of Nelson* (Allen and Unwin, 1962-); *John von Neumann: collected works* (Pergamon, 1961-3).

c) *Festschriften:* these dedicatory volumes of papers by colleagues or former students, etc, are not uncommonly met with, particularly in the field of academic science, and once again biographical information is usually included, frequently in the form of one of the papers, *eg, Horizons in biochemistry: Albert Szent-Gyorgi dedicatory volume* (Academic Press, 1962), *Progress in applied mechanics: the Prager anniversary volume* (Collier-Macmillan, 1963).

d) *Bibliographies:* in a similar way bibliographies of individual scientists and technologists can also furnish very substantial amounts of biographical data in the form of introduction or notes, *eg,* W R LeFanu *A bio-bibliography of Edward Jenner, 1749-1823* (Harvey and Blythe, 1951), Sir Geoffrey Keynes *A bibliography of the writings of Dr William Harvey, 1578-1657* (CUP, second edition 1953). And as was noted above (page 97) many bibliographies list books and papers *about* as well as *by* their man. Even *subject* bibliographies may contain biographical accounts of authors, *eg,* J Ferguson *Bibliotheca chemica* (Glasgow, Maclehose, 1906).

e) *Lists of award winners:* these vary considerably in the amount of biographical detail included. It is sparse in the Special Libraries Association *Handbook of scientific and technical awards in the United States & Canada, 1900-1952* (New York, 1956), but more extensive in E Farber *Nobel Prize winners in chemistry, 1901-1961* (Abelard-Schumann, revised edition 1963) and N H de V Heathcote *Nobel Prize winners in physics, 1901-1950* (Abelard-Schumann, 1953).

PORTRAITS

A portrait of a scientist or a technologist is an obvious example in the biographical field of a non-textual information source as described in the previous chapter. For the information a portrait conveys there is no substitute, short of an actual face to face meeting, and a number of the biographical sources that have been mentioned above contain portraits with the text, *eg,* Farber *Nobel Prize winners in chemistry, Biographical memoirs of Fellows of the Royal Society, Hackh's Chemical dictionary,* and most of the individual biographies, correspondence, collected works, festschriften, etc. There are examples of sources that make a prime feature of portraits: H M Smith *Torchbearers of chemistry* (New York, Academic Press, 1949) is a collection of over 200 portraits with very brief biographical sketches. Some tools, while not containing portraits themselves, do indicate where they can be

found, *eg,* Ireland *Index to scientists,* Williams *A biographical dictionary of scientists.* The field of medicine is uniquely fortunate to have
available the New York Academy of Medicine Library *Portrait catalog*
(Boston, Hall, 1960) in five volumes, comprising reproductions by
photo-litho offset of catalogue entries not only for over 150,000
portraits from books and journals, but also for over 10,000 separate
portraits (photographs, paintings, etc) in the academy's collections.

8

INDEX

No attempt has been made to index the many hundreds of titles mentioned in the text. As explained above (page 9), these are merely examples chosen from many possible alternatives as representatives of their class.

229

Technical journals 11, 107-8, 109, 142, 198, 202
Technical reports *see* Research reports
Technology 100
Terminology 40-2, 45, 184. *See also* Dictionaries
Tertiary sources of information 15, 20, 22, 69, 86, 87. *See also* individual types of sources, *eg* Directories
Textbooks 10, 11, 15, 21, 54, 55, 74, 77-85, 160, 170, 199, 200, 201
Thesauri 40, 125
Theses 14, 134, 175, 207-10, 211, 217
Titles 126-7, 137-8, 208
Trade associations 67, 200
Trade catalogues 199, 200
Trade directories 63-4, 66, 181, 200, 203
Trade journals 19, 69, 108-9, 181, 182
Trade literature 14, 107, 109, 198-206
Trade marks 179, 180-1, 182
Trade names 48, 64, 179-82

Translating dictionaries 43-7; of abbreviations 51
Translations 19, 190-7
Translators 191, 192, 193
Treatises 15, 18, 21, 30, 54-5, 71-4, 75, 76, 142, 158. *See also* Handbücher

Union lists *see* Location lists
Universities and colleges 16, 66, 104, 155, 165, 168, 191, 205. *See also* Theses
Unpublished sources of information 14, 93, 187, 192-4
User surveys 10, 15, 16-7, 18-9, 99, 111, 113, 114, 139-40, 149, 173, 190-1, 196-7

Video recording *see* EVR

Words *see* Dictionaries

Yearbooks 15, 20, 33, 70. *See also* Directories